SpringerBriefs in Molecular Science

Green Chemistry for Sustainability

Series editor

Sanjay K. Sharma, Jaipur, India

More information about this series at http://www.springer.com/series/10045

Willey Liew Yun Hsien

Towards Green Lubrication
in Machining

 Springer

Willey Liew Yun Hsien
Faculty of Engineering
Universiti Malaysia Sabah
Kota Kinabalu, Sabah
Malaysia

ISSN 2212-9898
SpringerBriefs in Molecular Science
ISBN 978-981-287-265-4   ISBN 978-981-287-266-1   (eBook)
DOI 10.1007/978-981-287-266-1

Library of Congress Control Number: 2014955794

Springer Singapore Heidelberg New York Dordrecht London

Springer Science+Business Media Singapore Pte Ltd. is part of Springer Science+Business Media
(www.springer.com)

*Dedicated to Tseng Chee Fung
and Cayden Liew Jin Wern*

Preface

Phosphorus, sulphur, zinc dialkyldithiophosphates (ZDDP) are examples of some of the widely used additives in lubricants. However, the concern for zinc and phosphorus as environmental contaminants as well as their poor biodegradability has resulted in efforts to find more environmentally benign replacements for industrial applications. Vegetable oils are viable and good alternative resources because of their environmental friendly, non-toxic and readily biodegradable nature. With the increasing cost associated with the procurement and disposal of traditional cutting fluids, and the threats on environmental and operator's health, alternative cutting fluids and lubrication methods are needed. The effectiveness of various types of vegetable oils as lubricants and additives in reducing wear and friction is discussed in this book. The book also provides information on the utilisation of environmental friendly gaseous and vapour, refrigerated compressed gas, chilled air, solid lubricant, mist lubrication and minimum quantity lubrication (MQL) in machining. Engineers and scientists working in the fields of lubrication and machining will find this book useful.

Contents

About the Author

Dr. Willey Liew Yun Hsien is Associate Professor at the Faculty of Engineering, Universiti Malaysia Sabah since August 2004. He received his B.E in Mechanical Engineering (1st class honours) from University of Leicester in 1991 and Ph.D. from University of Cambridge in 1998 under the Cambridge Commonwealth Trust and the Overseas Research Students Scholarship. He was the recipient of Institution of Mechanical Engineers (IMechE)/Shell Oils Tribology Award, Institution of Mechanical Engineers (IMechE) Project Prize and Institution of Mechanical Engineers (IMechE) Frederick Barnes Waldron Prize. His current research interests include tribology, precision machining, nanocoatings, novel carburization process and renewable energy. His works have been published in journals such as Wear, Tribology Letters, International Journal of Machining Science and Technology, and International Journal of Precision Engineering and Manufacturing. He serves as a reviewer for several international journals and conferences.

Abbreviations

ϕ Shear angle
σ Normal stress at the rake face
K Shear stress of the chip
μ Coefficient of friction at the rake face
τ Shear stress at the chip-tool interface
R Total force exerted by the tool
F Frictional force at the rake face
W Normal force at the rake face
T Cutting force at the rake face
N Thrust force at the rake face
α Rake angle at the rake face
β Mean angle of friction at the rake face

Chapter 1
Introduction

Abstract This chapter describes the friction and lubrication conditions at the chip–tool interface in a machining process, and provides a review of the research works carried out to investigate the mechanisms by which the lubricants penetrate into the chip–tool interface, and reduce the wear and cutting forces. This information is useful for researchers to formulate effective lubricants for machining processes.

Keywords Boundary lubrication · Hydrodynamic lubrication · Machining mechanism · Cutting tool wear

1.1 Lubrication Conditions

Lubrication separating surfaces can be characterised as hydrodynamic, boundary or mixed lubrication. In hydrodynamic lubrication, the contact surfaces are separated by a thin oil film. Boundary lubrication is a condition in which a lubricating film becomes too thin to provide total separation of the surfaces. It usually occurs under the combination of high-load and low-speed condition. This may also be due to a change in the characteristic of the fluid, resulting in contact between surface asperities. Under this lubrication condition, asperities on one surface collide with the asperities of the other sliding surface. The heat produced during sliding and collision induces chemical reactions between the nascent surfaces and the lubricant elements, resulting in the formation of lubricating film. The type of lubricating film formed on the contact surfaces has significant effect on the friction condition and the wear mechanism.

In mixed lubrication, both hydrodynamic and boundary lubrication prevails at the same time. The load, speed, fluid viscosity, temperature, surface roughness determine the type of lubrication to occur. The Stribeck curve in Fig. 1.1 illustrates the effect of friction, viscosity, load and sliding speed on the friction coefficient. The combination of low speed, low viscosity and high load will produce boundary

© The Author(s) 2015
W. Liew Yun Hsien, *Towards Green Lubrication in Machining*,
SpringerBriefs in Green Chemistry for Sustainability,
DOI 10.1007/978-981-287-266-1_1

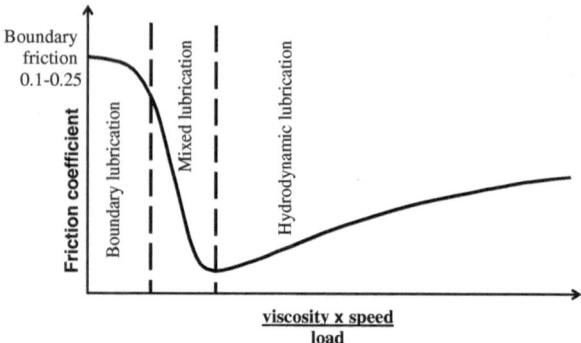

Fig. 1.1 Effect of viscosity, speed and load on coefficient of friction

lubrication, resulting in very high friction. As the speed and viscosity are increased, or the load is decreased, the surfaces will begin to separate by a fluid film. This results in the formation of mixed lubrication and hence a sharp drop in friction coefficient. The drop in friction is a result of decreasing surface contact due to the existence of a lubricating film at the contact surfaces. Separation of the surfaces increases as the speed is increased or the load is decreased. Eventually, the friction coefficient reaches to a minimum value and hydrodynamic lubrication is fully established. At this point, the load on the interface is entirely supported by the fluid film. Hydrodynamic lubrication also occurs in highly viscous liquid, and the lubricating characteristics depend on the properties of the lubricant as the load is fully supported by the lubricant. The friction and wear in hydrodynamic lubrication is low since the contact surfaces is fully separated by the lubricant and there is no contact surfaces. The increase in friction in the hydrodynamic region is due to fluid drag (friction produced by the fluid). Higher speed may result in thicker fluid film, but it also increases the fluid drag on the moving surfaces.

1.2 Mechanism of Machining

Machining is the most important secondary metal forming operation despite the fact that many aspects of the process are still imperfectly understood. In machining, chip formation takes place by a process of intense plastic shearing in a region known as the primary shear zone extending from the tip of the cutting tool to the free surface, as indicated in Fig. 1.2. The primary shear zone is often idealised as a plane called shear plane. The angle of inclination of the shear plane to the direction of cutting is called the shear angle ϕ. The chip has a freshly created and thus clean surface and, as it flows up the rake face of the tool, it is subject to a very high normal stress. Under these conditions, strong adhesion may occur between chip and tool giving rise to additional shear in the region of the chip adjacent to the tool surface known as the secondary shear zone. Friction between the rake face of a cutting tool and the freshly formed chip surface plays a vital role

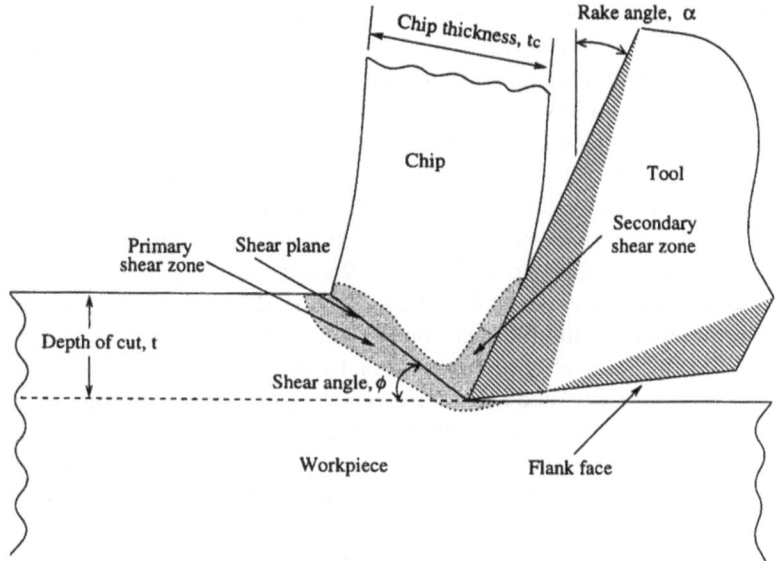

Fig. 1.2 The geometry of orthogonal cutting

in influencing both the ease of cutting and the quality of the resultant machined surface. The existence of clean surfaces together with the high local hydrostatic stresses favours the formation of strong adhesion between the cutting tool and the chip. These adhesive bonds can lead to poor surface integrity, although their extent can be limited by the provision of a suitable machining lubricant.

In addition to the primary and secondary zones, sliding of the tool flank face on the machined surface also generates heat which has significant effect on machinability of the material. The frictional sliding is accompanied by the bulk plastic deformation of work metal in the deformation process. High friction results in excessive metal flow in the vicinity of the interface resulting in poor surface finish, severe tool wear and high temperature. It is generally accepted that the most significant attribute of successful cutting lubricants are their ability to dissipate the heat at the cutting zone, and control the extent of the rake face adhesion and reduce frictional force at the contacted surface through the formation of effective boundary lubricants (Sect. 1.1). To be able to form such a lubricating film, the cutting lubricant must penetrate into the chip–tool interface.

1.3 Action of Lubricant in Reducing Friction in Machining

Research so far has shown that cutting fluids are needed to reduce the cutting resistance, improve the surface finish and prolong the tool life. Many research works have been carried out to investigate the issue pertaining to the penetration of the lubricants into the chip–tool interface and the mechanism by which the lubricants reduce the wear and cutting forces. This information is useful for researchers to formulate effective lubricants for machining. Much of the works utilised carbon tetrachloride as the test lubricant.

The diversity of opinions concerning the access of cutting fluids to the chip–tool interface may be illustrated by reviewing some of the hypotheses proposed. Cassin and Boothroyd (1965) suggested that the access of fluid was by diffusion through the distorted lattice structure, while it was undergoing severe plastic deformation in the primary zone. Barlow (1966–1967) using radioactively tagged carbon tetrachloride argued against this hypothesis and presented evidence to show that the effect of the lubricant on the emergence of dislocations at the free surface of the chip was relatively unimportant when the lubricant was available on the rake face region. The rake face lubrication proposed by Barlow did not exclude the possibility of another mechanism of penetration which had been put forward by Merchant (1957), who suggested that the cutting fluid was drawn to the chip–tool interface by the microcapillary action of the interlocking network of surface asperities.

More recently, a diffusion model had been developed to incorporate the transport of carbon tetrachloride through the fissures in the chip and the capillaries along the tool–chip interface, and thus the important of bulk diffusion through the fissures was still controversial (Smith et al. 1988). Hain (1952) estimated that capillaries as small as 2×10^{-7} m would be adequate to allow penetration of some fluids into the interface. This network had been identified at some distance away from the cutting edge where there was gross transfer of chip material onto the cutting tool (Dolye et al. 1979). Over the part of this contact, sliding friction (Sect. 4.1) occurred in which the coefficient of friction could be altered by the penetration of lubricants. The reduction in the tool–chip contact surface, cutting forces, cutting temperature and tool wear brought about by water vapour and gaseous (oxygen and carbon dioxide) in machining steel and titanium alloy were attributed to their ability to penetrate into the tool–chip interface through the capillaries (Liu et al. 2005, 2007; Wu and Han 2013).

Following penetration, the cutting fluid must be able to effectively modify the friction mechanism. It is generally accepted that the major lubricating action of the fluids at the chip–tool interface through the formation of a film tends to restrict metal-to-metal contact. The most significant attribute of successful cutting lubricants is their ability to penetrate into the chip–tool interface and reduce the extent of adhesion through the formation of a lubricating film. Merchant (1957) suggested that the friction reducing action was accomplished by chemical reaction with the nascent chip to produce a solid lubricant. Since then, several researchers had reported that the effectiveness of the cutting fluid was governed by the ability of the additives in the lubricant to form a low-friction protective film on the chip–tool interface (Mould et al. 1972; Shokrani et al. 2012). Cassin and Boothroyd (1965) demonstrated that the same argument could be applied to a series of chemically distinct cutting fluids. During their investigation, it was noted that the applied carbon tetrachloride fluid reduced the shear strength of the work material on the shear plane. Merchant's model had been subsequently been accepted and the process of molecular transport within the interfacial capillaries network has been considered to be of central importance in controlling the junction growth (Williams and Tabor 1977). They suggested that the major effect of fluids was not reducing the shear strength at the chip–tool interface but lowering the cutting forces through the reduction of the chip–tool contact area. On the other hand, Childs (1972) observed that the application of a cutting fluid reduced the elastic contact length and thus induced a steeper interface shear stress gradient. This resulted in a reduced frictional force.

Williams and Tabor (1977) have developed a model which classified the factor controlling the penetration of gaseous oxygen in terms of two distinct transport mechanisms (Knudsen and viscous flows). The transition from on mode to the other depends on the ratio of the mean free path of the vapour molecules to the characteristic capillary dimension. At higher pressures, the mean free path is small compared with the characteristic dimensions of the channel. Collisions between molecules are predominant and the flow can be characterised as viscous flow. Knudsen flow occurs at lower pressures where the flow is controlled by collisions between individual vapour molecules and the channel walls as the mean free path increases. This model, which is developed based on the argument of maintaining shear stress flow, shows that a reduction in the chip–tool contact length must take place when the shear strength at the chip–tool interface is reduced by the presence of oxygen.

Barlow (1966–1967) found that chemical reaction between the lubricant and the work material resulted in a change in the strain-hardening characteristics of the metal in the shear zone. It was proposed that during the cutting process, carbon tetrachloride eased the movements of dislocations by reacting chemically the atoms, resulting in a reduced strain-hardening rate and shear stress in the deformation zone. Naerheim et al. (1986) suggested that cutting lubricant could facilitate chip formation and curl, and reduce cutting forces by aiding the plastic deformation at the surface of the chip and in the capillaries within the chip. The physical effect of carbon tetrachloride in cutting was also reported by Usui et al. (1961) who noted that the cutting forces on copper could be reduced when the fluid was applied to the back face of the chip.

On the other hand, Shaw (1958–1959) argued that a high strength layer must be present at the interface to produce greater secondary strength and chip curl, lower chip–tool contact length and hence lower cutting forces. It became evident that the use of a single parameter such as the strength of the interface layer was inadequate to evaluate the effectiveness of lubricants; the chemical interaction between the tool and chip surfaces was important too. Strong interfacial bonds which were responsible for high frictional force could be established by the reaction of the specimen oxide with the counterface oxide, such as that occurring between a metal oxide and a sapphire tool to form a complex oxide (Pepper 1976). On the other hand, it had been found that the formation of a weak interfacial layer by the action of carbon tetrachloride vapour reduced the interaction between the chip and the tool surface (Wakabayashi 1993). Wakabayashi et al. (1995) showed that the transport of lubricant molecules to the tool–chip interface was unlikely always to be the rate-controlling step in gas phase lubrication in providing monolayer deposition of lubricant molecules on the nascent chip surface. A new concept of gas phase lubrication involving the reaction of lubricant molecules with the fresh chip surface was proposed. These authors found that when the cutting speed was increased, the lubricating activity could be maintained by increasing either the vapour pressure or the temperature at the chip–tool interface, which in turn increased the rate of formation of the lubricating film on the chip nascent surface.

References

Barlow PL (1966–1967) Influence of free surface environment on the shear zone in metal cutting. Proc Inst Mech Eng 181:687–702

Cassin C, Boothroyd G (1965) Lubricating action of cutting fluids. J Mech Eng Sci 7:67–81

Childs THC (1972) Rake face action of cutting lubricants. Proc Inst Mech Eng 186:717–727

Doyle ED, Horne JG, Tabor D (1979) Frictional interactions between chip and rake face in continuous chip formation. Proc R Soc London (Ser A) 366:173–183

Hain GM (1952) Measuring the cooling properties of cutting fluids. Trans ASME 74:1079–1080

Liu J, Han R, Sun Y (2005) Research on experiments and action mechanism with water vapour as coolant and lubricant in green cutting. J Mach Tools Manuf 45:687–694

Liu J, Han R, Zhang L, Guo H (2007) Study on lubricating characteristic and tool wear with water vapor as coolant and lubricant in green cutting. Wear 262:442–452

Merchant ME (1957) Cutting-fluid action and the wear of cutting tool. In: Lubrication and wear conference of institute mechanical engineering, London, pp 566–573

Mould RW, Silver HB, Syrett RJ (1972) Investigation of the activity of cutting oil additives. Wear 22:269–286

Naerheim Y, Smith T, Lan MS (1986) Experimental investigation of cutting fluid interaction in machining. J Tribol Trans ASME 108:364–367

Pepper SV (1976) Effect of absorbed films on friction of Al_2O_3-metal systems. J Appl Phys 47:2579–2583

Shaw MC (1958–1959) On the action of metal cutting fluids at low speeds. Wear 2:217–227

Shokrani A, Dhokia V, Newman ST (2012) Environmentally conscious machining of difficult-to-machine materials with regard to cutting fluids. Int J Mach Tools Manuf 57:83–101

Smith T, Naerheim Y, Lan MS (1988) Theoretical analysis of cutting fluid interaction in machining. Tribol Int 21:239–245

Usui E, Gujral A, Shaw MC (1961) An experimental study of the action of carbon tetrachloride in cutting and other processes involving plastic flow. Int J Mech Tool Dev Res 3:187–197

Wu J, Han RD (2013) Research on experiments with water vapour as coolant and lubricant in drilling TiAl4V. Ind Lubr Tribol 65:50–60

Wakabayashi T, Williams JA, Hutchings IM (1993) The action of gaseous lubricants in the orthogonal machining of an aluminium alloy by titanium nitride coated tools. Surf Coat Technol 57:183–189

Wakabayashi T, Williams JA, Hutchings IM (1995) The kinetics of gas-phase lubrication in the orthogonal machining of an aluminium alloy. Proc IMechE, J Eng Tribol 209:131–136

Williams JA, Tabor D (1977) The role of lubricants in machining. Wear 43:275–292

Chapter 2
Utilization of Vegetable Oil as Bio-lubricant and Additive

Abstract The environmental and toxicity issues of conventional lubricants as well as their rising cost lead to renewed interest in the development of environmental friendly oils as lubricants and industrial fluids. This chapter provides a review of the fundamental research works carried out on tribotesters to investigate the effectiveness of vegetable oil in suppressing wear and frictional force. Results obtained in these studies are useful to explain the mechanism by which the vegetable oils reduce friction and tool wear in machining (Chap. 3). Intensive review of the previous works shows that vegetable oils have high potential to be used as lubricant and additive to replace conventional lubricants and additives.

Keywords Bio-additives · Bio-lubricant · Vegetable oil lubricant · Tribotesters · Palm oil lubricant

2.1 Exploration for Environmental Friendly Lubricant Additives

Additives are widely used to improve the lubricant performance of base oil. Without additives, even the best base fluids are deficient in some features. The performance of a lubricant depends collectively on the base oil, additives and formulation. Phosphorus, sulphur, zinc dialkyldithiophosphates (ZDDP) are examples of some of the widely used additives. Sulphur-containing additives are probably the earliest known additive compounds in lubricants. In recent decades, it had attracted a considerable amount of research efforts to further explore their potential as effective anti-wear (AW) and extreme pressure (EP) additive (Zhang et al. 1999; Bhattacharya et al. 1995).

Fatty acids, alcohols, amines and esters are some of the AW additives used to produce a molecular film adhering to the surfaces by physical or chemical adsorption (Stachowiak and Batchelor 2005). The lubricant films are built up of orderly and closely packed arrays of molecular layers, with the polar head of the additive

© The Author(s) 2015
W. Liew Yun Hsien, *Towards Green Lubrication in Machining*,
SpringerBriefs in Green Chemistry for Sustainability,
DOI 10.1007/978-981-287-266-1_2

molecule anchored on the worn surface (Kenbeck and Bunemann 2009). There are also strong dipole interactions between the chains. The effectiveness of the lubricant depends greatly on the tenaciousness of the bond between the polar end group of the molecular chain and the metal surface where it adheres to (Tan et al. 2002).

Sulphur-, chlorine- and phosphorus-containing compounds are commonly used as EP additives to provide protection in EP condition (Canter 2007). These additives would form layers of iron compounds such as sulphides, chlorides and phosphates, respectively, through tribochemical reactions (Hsu and Gates 2005). The mechanism of lubrication which is influenced by these additive elements involves some chemical changes on the surface to form a surface protection film. This film is called boundary lubricating film or a tribofilm. The tribofilm plays a major role in determining the friction and wear of the tribological interaction. The morphology, integrity and mechanical properties of the tribofilms may vary depending on the properties of rubbing material as well as the type of lubricant additives used (Biswas 2000; Kim et al. 2010).

ZDDP was initially used as an antioxidant, but their excellent AW properties were quickly recognised and had been investigated intensively by many researchers. The AW function of ZDDP was attributed to its decomposed products that led to the formation of sacrificial reaction layers on the rubbing surfaces. A variety of ZDDP decomposition mechanisms and the associated chemistry of reaction film had been proposed by many researchers (Mosey et al. 2005; Fuller et al. 1997, 1998; Brancroft et al. 1997; Willermet et al. 1995; Spedding and Watkins 1982). However, the concern for the content of heavy metal zinc and phosphorus as environmental contaminants had resulted in efforts to find more environmentally benign replacements for industrial applications (Cardis et al. 1989). It was stated that even ashless sulphur-containing compounds do not necessarily have good ecotoxicological profiles for environmentally friendly lubricants. Environmentally friendly lubricants must also have high level of biodegradability (Habereder et al. 2009).

The environmental and toxicity issues of conventional lubricants as well as their rising cost related to a global shortage and their poor biodegradability led to renewed interest in the development of environmental friendly lubricants. Environmental legislation by OSHA and other international regulation authorities discourage the use of mineral oil-based lubricant and environmental-harmful additives. There has been increasing demand for green lubricants and lubricant additives in recent years. Vegetable oils are viable and good alternative resources because of their environmental friendly, non-toxic and readily biodegradable nature. The majority of bio-lubricants are based on esters. There are natural esters which are triglycerides of vegetable oils. Oleochemical esters of fatty acids such as diesters, polyolesters and complex esters are derived from sunflower, rapeseed, palm oil and coconut. Triglycerides of vegetable oils are more polar than petroleum-based oils, thus they have a higher affinity to metal (Suarez et al. 2010). Owing to this character, vegetable oils and their derivatives are suitable for lubrication applications. Conversely, their low thermo-oxidation stability, primarily due to the presence of bis-allylic protons is the main limitation (Fox and Stachowiak 2007; Becker and Knorr 1996). They also have poor corrosion resistance (Ohkawa et al. 1995). Some studies have also shown that most vegetable oils undergo cloudiness, precipitation, poor flow, and solidification upon long-term exposure to cold temperature (Rhee et al. 1995; Kassfeldt and Goran 1997). Erhan et al. (2006) have demonstrated that thermo-oxidative stability and cold flow property can be improved using a combination of proper blending of chemical additives, diluent and high-oleic vegetable oils. Another major obstacle is the cost of the bio-lubricants. A bio-lubricant costs somewhere between 30–40 % more compared to a conventional lubricant. Lubricant formulations for more environmentally benign are, therefore, being developed based on their benefits and limitations.

Oils with more polar groups (like carboxylic acids and esters) possess more sites to react and adsorb with metal surfaces to provide boundary lubrication effects (Stachowiak and Batchelor 2005). A lubricating film with strong bonding to the surface and adequate cohesive interaction among lubricant molecules can effectively reduce the friction and the amount of wear. To maintain a low friction and wear, the lubricating film has to withstand extremes of temperature variations, shear degradation and maintain excellent boundary lubricating properties through strong physical and chemical adsorption with the metal.

The additive molecules dissolved in the oil are attracted to the surfaces by adsorption forces governed by their polarity (Sharma et al. 2009; Kalin et al. 2006). It was found that the friction reduction effect increased with larger amount of adsorptive polar group of the additive in the base oil (Tohyama et al. 2009; Kurth et al. 2007; Adhvaryu et al. 2004). In fact, the polarity of both the base fluids and the additives is very important because each component of the mixture is competing for the metal surfaces reaction. A polar additive that would normally adsorb and desorb reversibly from a metal surface in a non-polar base fluid might have a much lower concentration on the surface than in a formulation that contains a high concentration of polar base fluid such as ester or vegetable oil (Rudnick 2009). Works by Suarez et al. (2010) shows that the wear performance of ZDDP additives contained in polar base fluid is better than that of ZDDP blended in a non-polar base fluid as smaller wear track width and larger load carrying capacity features were observed on the wear track produced in the former lubricant. Hsu et al. (1988) suggested that the first and the foremost of the dynamic and sequential competition in a solution that contains more polar groups is the preferential adsorption of the most polar molecules onto the surface at a particular temperature. However, the increase in the amount of polar compounds in the oil could reduce the adsorption of additives on the metal surface due to competitive adsorption, whereby the efficiency of AW additives could thus decrease (Studt 1989) and it could also provide corrosive effects (Hsu and Gates 2005; Jimenez and Bermudez 2007).

2.2 Wear and Friction Reduction by Vegetable Oil as Bio-lubricant and Additive

Vegetable oils are viable and good alternative resources because of their environmental friendly, non-toxic and readily biodegradable nature. The triacylglycerol structure with long fatty acid chains and presence of polar groups in the vegetable oils make them amphiphilic in character, therefore allowing them to be an excellent choice as lubricants and functional fluids. These triacylglycerol molecules in vegetable oils orient themselves with the polar end at the solid surface making a closed packed monomolecular or multimolecular layer resulting in a surface film that provides desirable qualities in a lubricant (Rudnick 2009). Other advantages include very low volatility due to the high molecular weight of the triglyceride molecule and excellent viscosity properties. Table 2.1 shows several type of vegetable-based lubricants developed for industry applications.

Vegetable oils may not suitable to be used as lubricants in their natural form due to their poor thermo-oxidation stability, low temperature behaviour and other tribochemical degrading processes that occur under severe conditions of temperature, pressure shear stress and environment (Fox and Stachowiak 2007). However, they can be used effectively as additives, in particular to improve the polarity behaviour of non-polar base fluid solutions, which would contribute to better tribological performance. In the past several researchers have investigated the effectiveness of methyl ester as additive in diesel. Sulek et al. (2010) found that the presence of fatty

Table 2.1 Several type of vegetable-based lubricants developed for industry applications (Shashidhara and Jayaram 2010)

Type of oil	Application
Canola oil	Hydraulic oils, tractor transmission fluids, metalworking fluids, food grade lubes, penetrating oils, chain bar lubes
Castor oil	Gear lubricants, greases
Coconut oil	Gas engine oils
Olive oil	Automotive lubricants
Palm oil	Rolling lubricant,-steel industry, grease
Rapeseed oil	Chain saw bar lubricants, air compressor-farm equipment, Biodegradable greases
Safflower oil	Light-coloured paints, diesel fuel, resins, enamels
Linseed oil	Coating, paints, lacquers, varnishes, stains
Soybean oil	Lubricants, biodiesel fuel, metal casting/working, printing inks, paints, coatings, soaps, shampoos, detergents, pesticides, disinfectants, plasticisers, hydraulic oil
Jojoba oil	Grease, cosmetic industry, lubricant applications
Crambe oil	Grease, intermediate chemicals, surfactants
Sunflower oil	Grease, diesel fuel substitutes
Cuphea oil	Cosmetics and motor oil
Tallow oil	Steam cylinder oils, soaps, cosmetics, lubricants, plastics

acid methyl ester derived from rapeseed oil in diesel fuel resulted in 20 % decrease in friction and twofold decrease in wear. Similarly, Sukjit and Dearn (2011) demonstrated that adding as little as 5 % of fatty acid methyl ester derived from rapeseed in diesel fuel could result in reduction in the wear scar diameter by 40 %.

Malaysia is often viewed as a country that evolved from dependence on tin and rubber to export-oriented manufacturing dominated by electronics assembly, but the commodity that made the country to the technological frontier is palm oil. Palm oil is now a major pillar of Malaysia's industrialization and it holds a considerable lead in global markets. To ensure a sustainable growth of palm oil industry in the country and remains competitive in the global market, palm oil industry in Malaysia in recent years has been shifting to palm oil product diversification from a conventional commercial cultivation as its main export focus until more years to come. Research and development effort, therefore, became more critical, in particular, to explore and develop new palm oil-based products for higher value added in the palm oil chain. Recently, it has been promoted as a biofuel feedstock in compression ignition engines (diesel engines). Palm oil methyl ester has an ester functional group which is a classic example of additive used for lubrication (Canter 2007). Characterization palm oil methyl ester can be carried out using fourier transform infrared spectroscopy (Liew et al. 2014). Palm oil methyl ester was produced from crude palm oil through transesterification process, whereby the triglyceride of palm oil was reacted with an alcohol in the presence of a catalyst as represented by general equation in Fig. 2.1. R_1, R_2 and R_3 represent the hydrocarbon chains of the fatty acid of the triglyceride. This reaction yielded esters and glycerol, which are then separated, in which glycerol being removed as by-product. The palm oil methyl ester was characterised using fourier transform infrared spectroscopy (FTIR). The FTIR spectra shows 1,750 and 1,150 cm^{-1} peaks (Fig. 2.2) that correspond to C=O and C–O esters (Taufiq-Yap et al. 2011).

Various investigations had shown that palm oil methyl ester additives improved the lubrication performance of the diesel base oil (Masjuki and Maleque 1996a, 1996b, 1997; Maleque et al. 2000). Masjuki and Maleque (1997) reported that adding 5 vol% of palm oil methyl ester in the base oil lubricant resulted in low wear rate of EN31 steel ball bearing. Palm oil methyl ester, converted from crude palm oil through transesterification, has very low sulphur content (0.002 wt%), and therefore is environmental friendly. Liew et al. (2014) found that in the

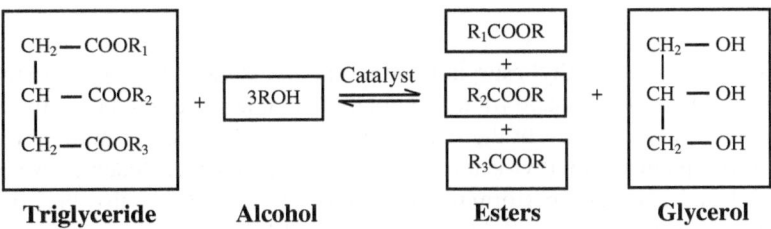

Fig. 2.1 Transesterification reaction for producing esters from oil (triglyceride)

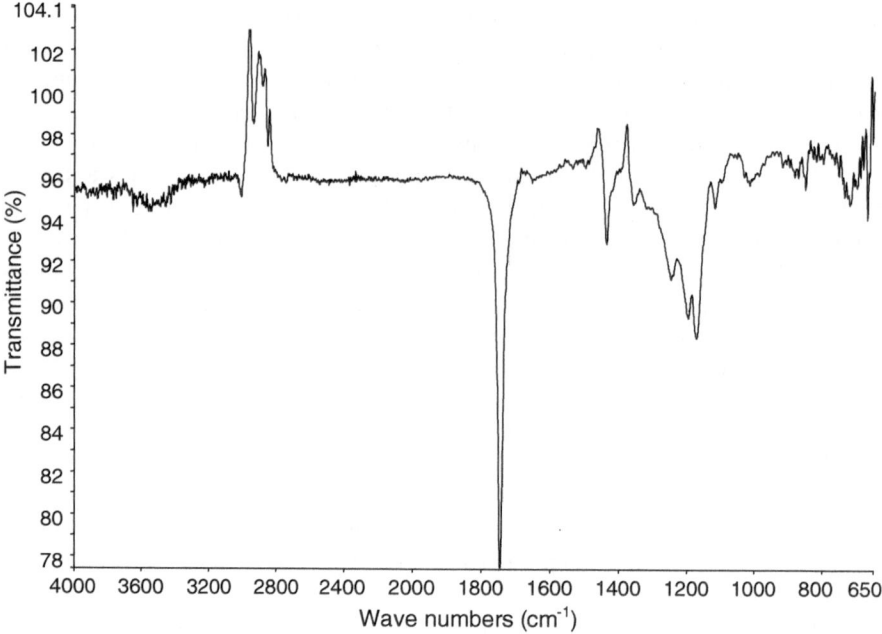

Fig. 2.2 The IR spectrum of palm oil methyl ester (Liew et al. 2014)

Fig. 2.3 The change in friction coefficient in different lubrication conditions at nominal load of 1,100 N (Liew et al. 2014)

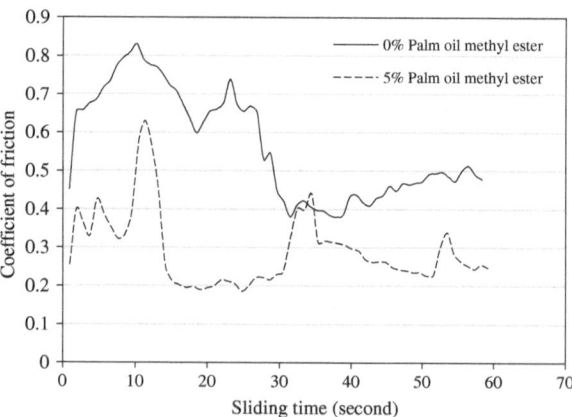

presence of palm oil methyl ester in the mineral oil resulted in a shorter running-in period and lower steady-state frictional coefficient at nominal load of between 600 and 800 N. The difference in the friction coefficient produced in mineral oil with and without palm oil methyl ester became more apparent at loads above 800 N (Figs. 2.3 and 2.4). The performances of lubricants in EP can also be expressed in terms of welding load (Kabuya and Bozet 1995; Singh and Verma 1991). Under mineral oil without palm oil methyl ester, complete welding of the four

Fig. 2.4 Effect of nominal load and lubrication condition on the average coefficient of friction (Liew et al. 2014)

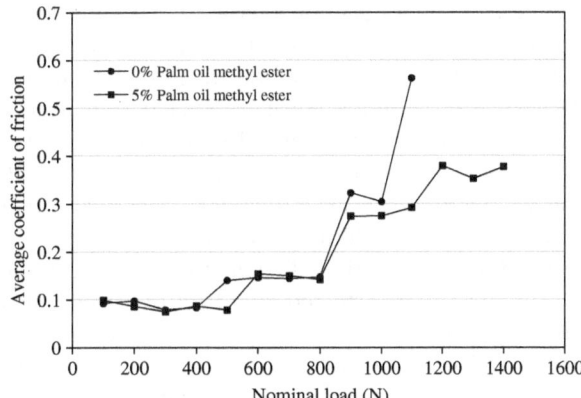

Table 2.2 Weld load and wear scar diameter for different lubrication condition

Lubrication condition	Weld load (N)	Average diameter scar (mm) produced at the nominal loads of		
		300 N	600 N	800 N
Mineral oil (without palm oil methyl ester)	1,200	0.28	1.97	2.60
Mineral oil (with 5 vol% palm oil methyl ester)	1,450	0.29	1.79	2.30

balls occurred at 1,200 N (Table 2.2). The presence of palm oil methyl ester in oil resulted in a higher critical load of 1,450 N.

Gong et al. (2003) investigated the wear reduction bought about by two kinds of synthetic thiophosphate (tri-*n*-octyl thiophosphate and tri-*n*-octyl tetrathiophosphate) and tricresyl phosphate as additives in rapeseed oil in sliding of steel. Synthetic thiophosphate resulted in lower wear and this could be attributed to the tribochemical reactions between the steel and the thiophosphate, and the formation of a boundary and protective layer on the worn surfaces. A series of long-chain dimercaptothiadiazole derivatives had been tested as AW and EP additives in vegetable oil using a four-ball tester. The long-chain thiadiazole derivatives were capable of improving the EP characteristic of the base colza oil. Thermal films generated from these derivatives are composed of ferrous sulphate and a small amount of adsorbed organic sulphide (Chen et al. 2012). Work by Gao et al. (1999) showed that thiadiazole derivatives in paraffin oil under boundary lubrication at high loads resulted in greater friction reduction and exhibited better AW properties than lubricant containing ZDDP. It was also found that thiadiazole derivatives had better antioxidative and anticorrosive properties than ZPPD.

Synthesis of vegetable oil and thiols could result in the formation of hydroxyl thioether derivatives in the vegetable oil. This process retained the vegetable oil structure and its associated benefits such as high flash point, viscosity index, lubricity and eco-friendly but removed poly-saturation in the fatty acid chain with addition of polar functional groups that significantly improved surface adsorption on metal leading to a reduction in wear and friction coefficient (Sharma et al. 2009). Xu et al. (2014a) demonstrated that catalytic esterification of crude bio-oil derived from spirulina algae resulted in enhanced lubrication performance. The coefficient of friction produced by ethanol blended with bio-oils esterified using potassium fluoride/alumina and potassium fluoride/HZSM-5 zeolite as catalysts was 22 and 10 % lower, respectively, than that produced by crude bio-oil blended with ethanol. The esterified bio-oils produced lower friction coefficient because it resulted in the formation of a better protective tribofilm on the worn surfaces.

Shi et al. (2014) studied the effect of water content in glycerol solution on the viscosity, friction coefficient, wear loss and film thickness under elastrohydrodynamic and boundary lubrication. Despite the viscosity and the film thickness of the glycerol solutions decreased greatly with increasing water content, there was an optimum amount of water in the glycerol solutions which resulted in the lowest friction coefficient. A correlation was not found between friction coefficient and wear volume loss. Under elastrohydrodynamic lubrication, the friction coefficient of rapeseed oil was about three times higher than that of pure glycerol. Under boundary lubrication, the friction coefficient produced by rapeseed oil and pure glycerol was similar. However, a more stable and lower friction coefficient was produced when the glycerol solution consisted of 5–20 wt% of water. While the water content was beneficial for glycerol to decrease the friction coefficient, excessive water content resulted in marked reduction in the viscosity and thus the load carrying capacity of the solution. These works showed that glycerol aqueous solutions have great potential to replace rapeseed oils as environmentally friendly base oils.

Although coconut oil is more stable than many vegetable oils, it is not widely used due to its high congelation temperature. Being a vegetable oil having typical triacylglycerol structure, it has most of the salutary properties of other vegetable oils as lubricants such as high viscosity index, good lubricity, high flash points and low evaporative loss. Although it has similar disadvantages of poor properties at low temperature, it shows much better thermal and oxidative stability owing to its predominantly saturated nature of its fatty acid constituents. Jayadas and Prabhakaran (2006) found that adding 2 wt% of ZDDP in coconut oil resulted in significant wear reduction in the four-ball test. The welding load of the coconut oil containing 2 wt% of ZDDP was higher than that produced by commercial 20W50 Oil.

Quinchia et al. (2014) studied the frictional and lubricating film-forming properties of various type of improved vegetable oils based lubricants (high-oleic sunflower, soybean and castor), using 4 wt% of ethylene–vinyl acetate copolymer and 1 wt% of ethyl cellulose as additives. It was found that castor oil showed the best lubricant properties, when compared to high-oleic sunflower and soybean oil, with very good film-forming properties and excellent friction and wear behaviour. This could be attributed to its hydroxyl functional group that increased both the viscosity and polarity of this vegetable oil. Ethylene–vinyl acetate copolymer exerted a slight effect on the lubricating film-forming properties, reducing the friction and wear mainly in the mixed lubrication region. Ethyl cellulose, on the other hand, was much more effective mainly with castor oil, in improving both mixed and boundary lubrication.

Alves et al. (2013) found that modified vegetable oils such as epoxidised sunflower and soybean oils resulted in lower friction coefficient than the mineral and synthetic oils. However, the presence of CuO and ZnO nanoparticles in the epoxidised vegetable oils resulted in higher friction coefficient and wear. It was postulated that the effect of nanoparticles on the wear and friction coefficient was governed by the nature of the adsorption of the lubricant on the contact surfaces. Adherence of the polar groups of the vegetable oils on the worn surface caused the nanoparticles to roll and hence three-body abrasion to take place, resulting in increased wear. The reduction in the wear and friction coefficient bought about these oxides nanoparticles in mineral and synthetic oils could be attributed to adherence of the nanoparticles and formation of a physical tribofilm on the worn surface. Xu et al. (2014b) reported that emulsified bio-oil (produced form the fast pyrolysis of rice husk) produced the lowest coefficient of friction. This was followed by bio-oil and diesel oil. The diesel and bio-oils produced the lowest and the highest wear, respectively. It was concluded that the emulsified bio-oil produced the best overall results and this was due the presence of various acidic components with polar groups in the emulsified bio-oil. Suarez et al. (2009) demonstrated that adding some soybean oil methyl esters and diesel-like pyrolytic fuel (produced through pyrolysis of soybean oil) enhanced the lubricity of diesel fuels. This result showed the potential use of both bio-fuels as additives for improving the tribological property of fossil fuels. Another work (Xu et al. 2010) found that the lubrication ability of the diesel fuel blended with bio-oil (produced through fast pyrolyzing rice husk) was better than that of the conventional diesel fuel. However, the presence of bio-oil in the diesel fuel resulted in inferior anti-corrosion and anti-wear properties.

References

Adhvaryu A, Erhan SZ, Perez JM (2004) Tribological studies of thermally and chemically modified vegetable oils for use as environmentally friendly lubricants. Wear 257:359–367

Alves SM, Barros BS, Trajano MF, Ribeiro KSB, Moura E (2013) Tribological behavior of vegetable oil-based lubricants with nanoparticles of oxides in boundary lubrication conditions. Tribol Int 65:28–36

Biswas SK (2000) Some mechanisms of tribofilm formation in metal/metal and ceramic/metal sliding interactions. Wear 245:178–189

Becker R, Knorr A (1996) An evaluation of antioxidants for vegetable oils at elevated temperatures. Lubr Sci 8:95–117

Bhattacharya A, Singh T, Verma VK, Prasad N (1995) 1,3,4-thiadiazoles as potential EP additives—a tribological evaluation using a four-ball test. Tribol Int 28:189–194

Brancroft GM, Kasrai M, Fuller M, Yin Z (1997) Mechanism of tribochemical film formation: stability of tribo- and thermally-generated ZDDP films. Tribol Lett 3:47–51

Canter N (2007) Special report: Trends in extreme pressure additives. Tribol Lubr Technol 63:10–18

Cardis AB, Davis RH, Piotrowski AB (1989) US Patent No. 4846984. U.S. Patent and Trademark Office, Washington, DC

Chen H, Yan J, Ren T, Zhao Y, Zheng L (2012) Tribological behavior of some long-chain dimercaptothiadiazole derivatives as multifunctional lubricant additives in vegetable oil and investigation of their tribochemistry using XANES. Tribol Lett 45:465–476

Erhan SZ, Sharma BK, Perez JM (2006) Oxidation and low temperature stability of vegetable oil-based lubricants. Ind Crops Prod 24:292–299

Fox NJ, Stachowiak GW (2007) Vegetable oil-based lubricants—a review of oxidation. Tribol Int 40:1035–1046

Fuller M, Yin Z, Kasrai M, Bancroft GM, Yamaguchi ES, Ryason PR, Willermet PA, Tan KH (1997) Chemical characterization of tribochemical and thermal films generated from neutral and basic ZDDPs using X-ray adsorption spectroscopy. Tribol Int 30:305–315

Fuller M, Kasrai M, Bancroft GM, Fyfe K, Tan KH (1998) Solution decomposition of zinc dialkyl dithiophosphate and its effect on antiwear and thermal film formation studied by X-ray absorption spectroscopy. Tribol Int 31:627–644

Gao Y, Zhang Z, Xue Q (1999) Study on 1,3,4-thiadiazole derivatives as novel multifunctional oil additives. Mater Res Bull 34:1867–1874

Gong Q, He W, Liu W (2003) The tribological behaviour of thiophosphates as additives in rape-seed oil. Tribol Int 36:733–738

Habereder T, Moore D and Lang M (2009) Eco requirements for lubricant additives. In: Rudnick LR (ed) Lubricant additives chemistry and applications. CRC Press Taylor & Francis Group, Florida

Hsu SM, Klaus EE, Chen HS (1988) A mechano-chemical descriptive model for wear under mixed lubrication conditions. Wear 128:307–323

Hsu SM, Gates RS (2005) Boundary lubricating films: formation and lubrication mechanism. Tribol Int 38:305–312

Jayadas NH, Prabhakaran KN (2006) Coconut oil base oil for industrial lubricants—evaluation and modification of thermal, oxidative and low temperature properties. Tribol Int 39:873–878

Jimenez AE, Bermudez MD (2007) Ionic liquids as lubricants for steel-aluminum contacts at low and elevated temperatures. Tribol Lett 26:53–60

Kabuya A, Bozet JL (1995) Comparative analysis of the lubricating power between a pure mineral oil and biodegradable oils of the same mean iso grade. Tribology Series 30:25–30

Kalin M, Vizintin J, Vercammen K, Barriga J, Arnsek A (2006) The lubrication of DLC coatings with mineral and biodegradable oils having different polar and saturation characteristics. Surf Coat Technol 200:4515–4522

Kassfeldt E, Goran D (1997) Environmentally adapted hydraulic oils. Wear 207:41–45

Kenbeck D, Bunemann TF (2009) Organic friction modifiers. Lubricant additives chemistry and applications. CRC Press Taylor & Francis Group, Florida

Kim BH, Mourhatch R, Aswath PB (2010) Properties of tribofilms formed with ashless dithi-ophosphate and zinc dialkyl dithiophosphate under extreme pressure conditions. Wear 268:579–591

Kurth TL, Jeffrey AB, Cermak SC, Sharma BK, Biresaw G (2007) Non-linear adsorption model-ling of fatty esters and oleic estolide esters via boundary lubrication coefficient of friction measurements. Wear 262:536–544

Liew WYH, Dayou S, Dayou J, Siambun NJ, Ismail MAB (2014) The effectiveness of palm oil methyl ester as lubricant additive in milling and four-ball tests. Int J Surf Sci Eng 8:153–172

Masjuki HH, Maleque MA (1996a) The effect of palm oil diesel fuel contaminated lubricant on sliding wear of cast irons against mild steel. Wear 198:293–299

Masjuki HH, Maleque MA (1996b) Wear, performance and emissions of a two-stroke engine running on palm oil methyl ester (POME) blended lubricant. Proc Part J IMechE J Eng Tribol 210:213–219

Masjuki HH, Maleque MA (1997) Investigation of the anti-wear characteristics of palm oil methyl ester using a four-ball tribometer test. Wear 206:179–186

Maleque MA, Masjuki HH, Haseeb ASMA (2000) Effect of mechanical factors on tribological properties of palm oil methyl ester blended lubricant. Wear 239:117–125

Mosey JM, Muser MH, Woo TK (2005) Molecular mechanisms for the functionality of lubricant additives. Science 307:1612–1615

Ohkawa SA, Konishi H, Hatano K, Tanaka K, Iwamura M (1995) Oxidation and corrosion char-acteristics of vegetable base biodegradable hydraulic oils. SAE Tech Paper 951038:55–63

Quinchia LA, Delgado MA, Reddyhoff T, Gallegos C, Spikes HA (2014) Tribological studies of potential vegetable oil-based lubricants containing environmentally friendly viscosity modifiers. Tribol Int 69:110–117

Rhee IS, Valez C, Bernewitz K (1995) Evaluation of environmentally adapted hydraulic fluids, TARDEC tech report 13640. US army tank-automotive command research, Development and Engineering Center, Warren, MI, pp 1–15

Rudnick LR (2009) Additives for industrial lubricant applications. Lubricant Additives Chemistry and Applications. CRC Press Taylor & Francis Group, Florida

Sharma BK, Adhvaryu A, Erhan SZ (2009) Friction and wear behavior of thioether hydroxyl vegetable oil. Wear 42:353–358

Shashidhara YM, Jayaram SR (2010) Vegetable oil as a potential cutting fluid—an evolution. Tribol Int 43:1073–1081

Shi Y, Ichiro M, Mattias G, Marcus B, Roland L (2014) Boundary and elastohydrodynamic lubrication studies of glycerol aqueous solution as green lubricants. Tribol Int 69:39–45

Singh T, Verma VK (1991) EP activity evaluation of tris(N-arylthiosemicarbazido)-molybdenum(III) on steel balls in a four-ball test. Wear 146:313–323

Spedding H, Watkins RC (1982) Antiwear mechanism of ZDDPs—part 1. Tribol Int 15:9–12

Stachowiak GW, Batchelor AW (2005) Engineering tribology, 3rd edn. Elsevier, Oxford

Studt P (1989) Boundary lubrication: adsorption of oil additives on steel and ceramic surfaces and its influence on friction and wear. Tribol Int 22:111–119

Suarez AN, Grahn M, Pasaribu R, Larsson R (2010) The influence of base oil polarity on the tribological performance of zinc dialkyl dithiophosphate additives. Tribol Int 43:2268–2278

Suarez PAZ, Moser BR, Sharma BK, Erhan SZ (2009) Comparing the lubricity of biofuels obtained from pyrolysis and alcoholysis of soybean oil and their blends with petroleum diesel. Fuel 88:1143–1147

Sukjit E, Dearn KD (2011) Enhancing the lubricity of an environmentally friendly Swedish diesel fuel MK1. Wear 271:1772–1777

Sulek MW, Kulczycki A, Malysa A (2010) Assessment of lubricity of compositions of fuel oil with biocomponents derived from rape-seed. Wear 268:104–108

Tan Y, Huang W, Wang X (2002) Molecular orbital indexes criteria for friction modifiers in boundary lubrication. Tribol Int 35:381–384

Taufiq-Yap YH, Abdullah NF, Basri M (2011) Biodiesel production via transesterification of palm oil using HaOH/Al$_2$O$_3$ catalysts. Sains Malaysia 40:587–594

Tohyama M, Ohmori T, Murase A, Masuko M (2009) Friction reducing effect of multiply adsorptive organic polymer. Tribol Int 42:926–933

Willermet PA, Dailey DP, Carter RO III, Schmitz PJ, Zhu W (1995) Mechanism of formation of antiwear films from zinc dialkyldithiophosphates. Tribol Int 28:177–187

Xu Y, Wang Q, Hu X, Li C, Zhu X (2010) Characterization of the lubricity of bio-oil/diesel fuel blends by high frequency reciprocating test rig. Energy 35:283–287

Xu Y, Zheng X, Hu X, Dearn KD, Xu H (2014a) Effect of catalytic esterification on the friction and wear performance of bio-oil. Wear 311:93–100

Xu Y, Zheng X, Yin Y, Huang J, Hu X (2014b) Comparison and analysis of the influence of test conditions on the tribological properties of emulsified bio-oil. Tribol Lett 55:543–552

Zhang J, Liu W, Xue Q, Ren T (1999) A study of N and S heterocyclic compound as potential lubricating oil additive. Wear 224:160–164

Chapter 3
Utilisation of Vegetable Oil, Solid Lubricant, Mist Lubrication, Minimum Quantity Lubrication (MQL) in Machining

Abstract Cutting fluids are widely used in most machining processes to produce superior surface finish and suppress tool wear. Due to the increasing cost associated with the procurement and disposal, and environmental issues of traditional cutting fluids, researches have been carried out to search for biodegradable lubricants and alternative lubrication methods to reduce the dependency on traditional cutting fluids. The growing demand for biodegradable lubricants has opened an avenue for using vegetable oils as alternative lubricants for machining. This chapter discusses the effectiveness of various types of biodegradable oils as lubricants and additives, and alternative lubrication methods such as minimum quantity lubrication (MQL), mist lubrication and solid lubrication in machining processes.

Keywords Cutting fluids · Biodegradable lubricants · Alternative lubrication methods · Tool wear reduction

3.1 Tool Wear in Machining

Tool wear plays a vital role in influencing both the ease of cutting and the quality of the resultant machined surface. In machining, the cutting parameter and environment had significant influence on the tool wear mechanism and surface finish, and hence the quality of the products. Abrasion and attrition wear are the dominant tool wear mechanisms in low-speed machining. During low-speed machining, unstable built-up edge (BUE) tends to form (Korkut and Donertas 2007).

Generally, BUE forms when shearing takes place at some particularly weak point within the chip itself because the bonding force between the strain hardened chip and the tool exceeds the shear strength of the main body of the chip. The presence of the BUE is sometimes beneficial because it increases the rake face angle of the tool and hence reduces the cutting forces forces (Rowe 1981; Shaw 1984). However, the formation of a BUE is not normally preferable because it often results in a deterioration of the surface finish. The BUE tends to protrude below

© The Author(s) 2015
W. Liew Yun Hsien, *Towards Green Lubrication in Machining*,
SpringerBriefs in Green Chemistry for Sustainability,
DOI 10.1007/978-981-287-266-1_3

the level of the tool nose and prevent contact between the tool land and freshly machined workpiece surface. As the BUE grows forward, it will usually also extend downward causing the finished surface to be undercut. BUE breaks down when it becomes unstable. Broken pieces may adhere to the chip. Those adhering on the underside of the machined surface cause deterioration in the surface finish. Fracture of the nose region can result in large cavities and broken step-like flaws (Elkhabeery and Bailey 1984). Grooves and surface deformation may also be produced by the ploughing action of the BUE. When BUE is detached from the tool, it can cause particles to be plucked out from the tool, resulting in the formation of cavities. This wear mechanism is known as attrition (Gu et al. 1999; Korkut and Donertas 2007; Liew and Ding 2008).

The flow of the chips is less laminar in the low-speed machining and this can cause extensive attrition wear (Liew and Ding 2008). Gu et al. (1999) found that attrition wear was the dominant tool wear mechanism in milling steel at low speeds. During low-speed machining, the abrasive wear is also likely to take place as the work material is hard enough (due to low heat generated) to plough into the tool (Gu et al. 1999; Liew and Ding 2008). Abrasive wear results in the formation of grooves (Kumar et al. 2006; Liew and Ding 2008; Lim et al. 1999; Marinov 1996). In ultra-precision machining of stainless steel using polycrystalline cubic boron nitride (PCBN) tools at low speeds (less than 50 m/min), fine-scale cavities were formed on the rake face in the initial stage of machining (Fig. 3.1). The damaged surface acted like a chip breaker and thus as a preferential site for crack initiation. Once a crack was initiated, it propagated along the grain boundaries leading to intergranular fracture (Liew et al. 2003).

Increasing the cutting speed and feed rate can result in the elimination of the BUE, leading to a reduction in attrition wear (Korkut and Donertas 2007; Abou-El-Hossien and Yahya 2005). As the cutting speed is increased, the chip-tool interface temperature increases and the BUE disappears when thermal softening at the interface causes a lower flow stress at the interface than in the main body of

Fig. 3.1 SEM of the rake faces of the PCBN tools grade BN 100 used to machine stainless steel for a distance of 1,320 m at 44 m/min at a depth of cut of 20 μm (Liew et al. 2003)

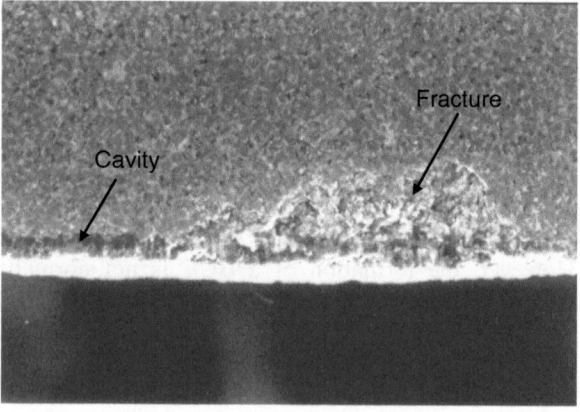

the chip. High-speed machining offers many advantages such as higher material removal rate, and better surface finish due to the elimination of BUE (Liao and Lin 2007; Su et al. 2006). However, milling at high speeds can cause short tool life (Chakraborty et al. 2008; Gu et al. 1999; Kim et al. 2001; Soderberg and Hogmark 1986; Vivancos et al. 2005) as it generates high temperature and thus thermal cracking and thermo-chemical wear (such as diffusion and oxidation) (Dolinsek et al. 2001; Liao and Lin 2007; Liu et al. 2005; Nouari and Molinari 2005; Sokovic et al. 2004; Sun et al. 1998). Diffusion can cause a reduction in the strength of the surface layer and thus the tool edge more susceptible to fracture. On the other hand, the diffusion of atom from the tool to the workpiece can cause the workpiece to harden and thus the severity of abrasion. Nouari and Ginting (2006) reported that high cutting temperature generated in the high-speed milling caused diffusion at the cutting interface. Liao and Lin (2007) found that diffusion weakened the bonding of the binder in the carbide tool. On the other hand, Sun et al. (1998) reported that diffusion and recrystallisation increased the adhesive strength between the work material and the tool. This in turn caused the work material becoming less likely to be detached from the tool. As a result of this, the severity of the attrition wear was reduced.

In interrupted cutting process such as milling, the tool is heated during cutting and cooled when it leaves the cutting zone. Temperature variation can cause periodic expansion and contraction of the tools leading to the formation of thermal cracks which is also known as comb cracks. Comb cracks propagate in a direction perpendicular to the tool edge. The use of coolant can lead to an increase in the thermal variation and therefore make comb cracks more likely to form. Thermal cracks are more likely to form at high speeds since the amplitude of the temperature variation increases with increasing speed (Viera et al. 2001; Bhatia et al. 1980). It was reported that comb cracks were formed at cutting speed higher than 100 m/min. Below this cutting speed, no comb cracks were formed, but attrition and chipping took place due to the mechanical loading and unstable nature of BUE (Bhatia et al. 1978; Bhatia et al. 1980; Liu et al. 2005; Nordin et al. 2000). Thermal cracks can weaken the tool edge and thus make the tool edge more susceptible to chipping and increase the flank wear leading to a deterioration of the surface finish (Ghani et al. 2004; Gu et al. 1999; Liao and Lin 2007; Liao et al. 2007; Liu et al. 2005). Nordin et al. (2000) proposed that comb cracks are formed during the cooling stage of the cutting cycle. Compressive stresses developed during the heating stage cause plastic deformation at the tool edge. The subsequent cooling stage causes the surface of the tool cools down more rapidly than the inner part, this causes the contraction of the tool material and high tensile stresses developed at the tool surface. When this stress exceeds the local fracture stress, comb cracks will form. Coating can effectively improve the tool life in milling steel. Coating was found to improve the oxidation resistance of the tool, protect the tool against diffusion wear, enhance the lubricity of the tool and reduce the temperature variation in the tool, rendering it less susceptible to crack (Ghani et al. 2004; Liu et al. 2005; Nordin et al. 2000; Arndt and Kacsich 2003; Endrino et al. 2006).

3.2 Flood Lubrication in Machining

Cutting fluids or lubricants are formed by a complex mixture of chemicals. The chemical species in this mixture depends on various factors including the manufacturer and also the cooling and lubrication requirement of the machining process. They reduce friction between the cutting tool and the work surface, reduce wear, protect surface characteristics, reduce surface adhesion or welding, carry away generated heat, and flush away swarf, chips, fines and residues. Cutting fluids are designed for use in various machining operations such as turning, grinding, boring, tapping, threading, gear shaping, reaming, milling, broaching, drilling, hobbing and sawing. Cutting fluids can be manually applied in a flood form to the cutting zone of the tool and the work or delivered as a mist in a high-velocity air stream. A continuous stream of cutting fluids delivered by a low-pressure pump can be directed through a nozzle to the cutting edge of the machine tool or through the tool and over the work to carry away metal chips or swarf. A variety of fluid nozzle designs are available, depending on the application needed. A distribution system may be used to control cutting fluid flow volume and flow pressure.

Cutting fluids are broadly classified into two groups: oil based and water based. Water-based fluids can be further subgrouped into emulsifiable oils (soluble oils), synthetic and semi-synthetic fluids. Soluble oils contains a high percentage of oil (usually greater than 50 %) with some additives and emulsifiers. The high-oil content provides excellent physical lubricity for the cutting operation as well as corrosion protection for the machine tool. The water content provides cooling effect. Synthetic fluids, generally considered a coolant with low lubrication characteristics, are mineral oil free cutting fluids which contain inorganic or organic chemical solutions to provide water softening, corrosion prevention and lubrication. Semi-synthetic fluids contain both mineral oils and chemical additives and have both characteristics of soluble and synthetic fluids. The oil-based cutting fluids are usually mineral oils which often contain the extreme pressure additives such as chlorine, sulphates and phosphates (Shokrani et al. 2012). Chlorinated and sulphurised additives in extreme pressure cutting fluids chemically react with nascent surface of the chip and the tool resulting in formation of a low friction protective film. Minfray et al. (2014) investigated the performance of non-active paraffinic base oil with and without organic pentasulphide as EP additive in milling steel. It was found that the formation of iron sulphides (FeS and FeS_2) on the flank face of the cutter mill and on the chip contact surface due to the reaction between the iron and sulphur compounds in the additives resulted in reduction in wear and cutting force. In high-speed machining process, the excessive heat generated is the primary concern, whereby such conditions capable of damaging tools, product quality and other undesirable results. Under such circumstances, water-based fluids are likely to perform better than oil-based fluids.

General assessment of the machined surface was normally obtained by studying the effect of different process parameters and tool wear on the surface finish, waviness and dimensional accuracy. Many studies have shown that the machined subsurface condition has significant effect on the reliability of the machined components. Liew et al. (2004a, b) had studied the effectiveness of emulsified water-based

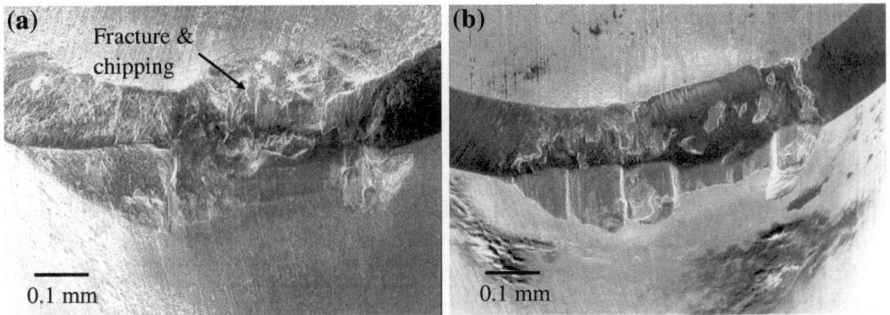

Fig. 3.2 SEM of the PCBN tools used to machine stainless steel with a hardness of 40 HRC for a distance of 5.65 km under **a** dry condition **b** flood lubrication using emulsified water-based coolant. Fracture and chipping was evident on the tools used to cut under dry condition (Liew et al. (2004a, b)

coolant (5–9 % concentration) in conventional turning of stainless steel using PCBN tools. Although the coolant could not bring about a reduction in the flank wear, it was effective in reducing the formation of white layer (untempered martensite), reducing the depth of the subsurface defect and in preventing chipping of the tool edge (Fig. 3.2), leading to improved surface finish. A white layer may form when the cutting temperature exceeds the austenisation temperature (to cause phase transformation of ferrite to austenite) followed by a rapid cooling. Overtempering (which may result in a reduction in the hardness) was also found to take place below the machined surface where the heat was not high enough to cause phase transformation. The lower cutting temperature when machining with coolant was indicated by the lower depth of affected sublayer and the absence of the white layer. Similarly, Ezugwu and Olajire (2002) had made a similar study on the effect of cutting fluid on the surface integrity in machining stainless steel. They found that the heat generated resulted in a reduction in the yield strength of the machined subsurface layers, hence emphasising the important of application of coolant during machining. Examination of the surface machined under dry condition revealed the existence of white layer.

Cutting fluids are used primarily to decrease the temperature in the cutting zone and lubricate the cutting interface in machining at high speeds (Kim et al. 2001; Reddy and Rao 2006). At low-speed machining, cutting fluid primarily functions as lubricant (Liao et al. 2007; Liew and Ding 2008). Therefore, cutting fluids with high lubricating ability are normally used in low-speed machining and cutting fluids with high cooling ability are used in high-speed machining. Emulsified water-based coolant flooded to the cutting zone was the conventional lubrication used in metal cutting. The tool life and surface finish can be improved by delivering cutting fluid into the cutting zone where the tool and the workpiece are in contact (Kim et al. 2001; Rahman et al. 2000; Stanford et al. 2007). The coolants can be sprayed at high pressure to increase the penetration of the cutting fluid into the cutting zone. Rahman et al. (2000) found that high-pressure coolant was effective in improving the tool life and the surface finish due to the elimination of the BUE. Similarly, Ezugwu and Bonney (2005) found that in machining Inconel 718, there was a critical coolant pressure under which the cutting tools had the highest tool life. The highest improvement in tool life (349 %) was achieved when machining at 60 m/min using coolant supplied at 11 MPa. Kaminski and Alvelid (2000) demonstrated that coolant applied at 20 MPa resulted in a marked reduction in the cutting temperature, i.e. 40 % lower than the temperature produced under flood lubrication.

However, some adverse effects arise with the application of cutting fluid in the machining process. Several researchers found that in milling steel at high speed, coolant promoted comb cracks (Kim et al. 2001; Liao and Lin 2007; Liao et al. 2007; Viera et al. 2001). Hence, the application of cutting fluid is not recommended in milling where comb cracking is predominant (Viera et al. 2001). The use of coolant increases the thermal variation and this causes thermal cracks more likely to form (Kim et al. 2001; Liew and Ding 2008; Viera et al. 2001). Vieira et al. (2001) had investigated the performance of various cutting fluid during face milling of steel. They found that when the cutting speed exceeded 110 m/min, the semi-synthetic cutting fluid exhibited better cooling ability, followed by the emulsion-based mineral oil, and the synthetic fluids. Higher degree of cooling caused greater temperature variation in the cutting tool and thus comb crack more likely to take place. Stanford et al. (2007) found that semi-synthetic coolant promoted chemical and corrosive wear leading to an increase in tool wear.

With the increasing cost associated with the procurement and disposal of traditional cutting fluids (generally termed as coolant), dry and semi-dry machining became increasingly attractive to the metal cutting industry. Most cutting fluids used threatens the operator's health, causing environment pollution, and increases the production cost (Rahman et al. 2002; Reddy and Rao 2006). All these problems led to the invention of alternatives cutting fluids and lubrication methods to reduce the usage of conventional flood coolants, overcome the problems arose from the conventional cutting fluid application, increase the tool life and improve the surface quality of the work. One way to eliminate totally the usage of lubricant is by carrying out the machining under dry condition.

While some alloys (in particular hardened steel) can be machined with PCBN tools to superior finish and form under dry condition, many other alloys showed poor machinability in the absence of lubricant (Yallese et al. 2009; Chou and Evans 1997; Liew et al. 2004a, b). There is a limited range of speed and materials that dry machining results in acceptable surface finish and tool life. In high-speed machining under dry condition, the intense heat generated at the cutting zone is the primarily concern because it can cause severe tool wear and surface deterioration. The same problem can also take place in machining alloy with low thermal conductivity (such as Inconel) even at moderate cutting speeds. In these circumstances, coolant is needed to dissipate the heat from the cutting zone. Machining soft materials such as aluminium and copper in the absence of lubricant is likely to produce poor surface finish due to the formation of BUE. Cutting fluids also flush the chips from the tool-workpiece interface. Without cutting fluids, the chips may entangle with the tool and the workpiece, deteriorating the machined surface. Deposition of solid lubricant on the cutting tool, MQL, chilled air lubrication and applying biodegradable lubricants are the alternatives lubrication methods to reduce the dependency on flood coolants.

3.3 Solid Lubricant

Sometimes, various types of lubrication methods (surface coating and lubrication) are used together to improve the machinability of materials. The coatings can be tailored in such a way that some of the phases can favourably react with additives in oil to provide much lower friction, and another phase can provide superhardness and hence low wear. Extensive researches have been pursued in this subject recently (Grill 1993; Haque et al. 2010; Kubo et al. 2008; Erdemir 2005; Mistry et al. 2011). Klocke et al. (2005) demonstrated that in machining steel alloys and Inconel, the tool life can be optimised by using the right combination of PVD-coated tool and additive-free synthetic esters as lubricant and coolant. Molybdenum disulphide solid lubricants filled into the micro-holes in the cemented carbide tools resulted in the formation of self-lubricating film at the tool–chip interface in dry machining of hardened steel (Deng et al. 2011). This, in turn, caused a marked reduction in the cutting forces, cutting temperature and roughness of the machined surface. It was postulated that the micro-holes containing molybdenum disulphide acted as a source of lubricant. It continuously released molybdenum disulphide to the rake face to maintain the lubrication at the tool-chip interface. More recently, Deng et al. (2012) studied the performance of conventional WC/Co carbide tools and textured WC/Co carbide tools filled with molybdenum disulphide in dry machining of steel. They found that the cutting forces, cutting temperature and the friction coefficient produced by the former tools were significantly lower. This could be due to the combination of a smaller contact surface on the textured rake face and the formation of a lubricating film on the tool-chip interface. Reddy and Rao (2006) investigated the effect of graphite and molybdenum disulphide solid lubricants deposited on the cutting tools on the surface finish, cutting forces and specific energy in milling AISI 1045 steel. The results showed that, compared to application of coolant, solid lubricant was more effective in improving the tool life and the quality of the surface. Broniszewski et al. (2013) investigated dry machining of hardened steel with Al_2O_3 tools containing various percentage of molybdenum (5–20 wt%) and found that the optimum percentage of molybdenum that gave the highest tool life was 15 wt%.

3.4 Mist and Minimum Quantity Lubrication (MQL)

Chakraborty et al. (2008) demonstrated that synthetic coolant sprayed in mist form was effective in resisting flank wear progression in milling of steel at 183 m/min. Results obtained in earlier work by Liew et al. (2003, 2004b), Ding et al. (2002) showed that in turning of stainless steel at low speeds with carbide and PCBN tools where the tool wear was governed by fracture and abrasive wear, liquid paraffin oil and cyclomethicone sprayed in small quantity in mist form was effective in reducing the tool wear and improving the surface finish.

Ko et al. (1999) applied air-oil mist lubrication in turning hardened steel. They reported that this method was more effective than flood coolant in reducing tool wear. The improved tool performance could be attributed to the simultaneous reduction in the temperature and friction by the mist coolant. Su et al. (2006) had conducted high-speed end milling of Ti-6Al-4V titanium alloy under various cooling and lubrication conditions to determine the optimal cooling and lubrication condition for the longest tool life. Dry, flood coolant, nitrogen-oil mist, compressed cold nitrogen gas, and mixture of compressed cold nitrogen gas and oil mist were used. The experimental results showed that mixture of compressed cold nitrogen gas and oil mist provided the best tool life among all the cooling and lubrication conditions used. The tool life obtained under compressed cold nitrogen gas and oil mist lubrication was 2.69 times of that obtained under dry cutting condition and 1.93 times higher than that obtained under nitrogen-oil-mist.

MQL is a technique where a very small amount of lubricant, 6–100 ml/h, is delivered to the tool cutting zone using compressed air stream at high pressure (Attanasio et al. 2006; Kang et al. 2008; Liao and Lin 2007; Liao et al. 2007; Rahman et al. 2002). The removal of heat generated during cutting is achieved mainly by the convection of the compressed air and partially by evaporation of the lubricant. Rahman et al. (2002) found that MQL was very effective in improving the tool life and surface finish in milling at low speed, feed rate and depth of cut. Similarly, Kang et al. (2008) found that MQL improved the cutting performance and tool life in high-speed end milling of AISI D2 die steel. Attanasio et al. (2006) reported that MQL could lubricate the tool–chip interface without inducing thermal shock and consequently reduced the probability of tool failure due to comb cracks.

Liao et al. (2007) reported that MQL improved the tool life and surface finish in both low- and high-speed milling up to 250 m/min. MQL could reduce the cutting temperature at high-speed milling and bring lubrication effect at low-speed milling. They also had found that different viscosity of the oil used in MQL resulted in different tool life. At low-speed milling (below 150 m/min), high viscous oil resulted in longer tool life than low viscous oil, while opposite results were obtained in high-speed milling. This could be due to the fact that low-viscosity oil volatilised more easily (as it contained higher fraction of low molecular weight components) and hence provided better cooling effect at high-speed machining. Many researchers had studied the effectiveness of vegetable oils as MQL lubricants in machining. This is discussed in detail in Sect. 3.5.

3.5 Vegetable Oil as Additive and Lubricant in Machining

Rahim and Sasahara (2011) evaluated the effect of MQL of palm oil and synthetic ester on the tool wear, cutting temperature and cutting force in high-speed drilling of titanium alloys. It was found that the palm oil produced lower flank wear than the synthetic ester and the conventional flood lubrication. Palm oil produced lower cutting forces and temperature than synthetic ester, but similar magnitude forces and temperature as produced by the flood lubrication. In machining low alloy steel AISI 9310, vegetable oil applied using MQL technique resulted in lower chip–tool interface temperature, flank wear and better surface finish, compared to the results produced by conventional flood lubrication (Khan et al. 2009). Similarly, Sharif et al. (2009) compared the effectiveness of palm oil sprayed using MQL technique and conventional flood lubrication in end milling of AISI 420 stainless steel. Palm oil sprayed using MQL technique produced 4 times higher tool life.

In machining aluminium silicon alloy, MQL of rapeseed oil brought about a small lubricating effect (Itaogawa et al. 2006). The boundary film developed on a tool surface was not strong enough to sustain low friction and to reduce adhesion of work material. The specific and axial force produced was found to be higher than those produced under flood lubrication. However, under the circumstances where mixture of oil and water droplets was used as MQL lubricant, a much lower forces were obtained, indicating that a cooling effect was very important to produce a low and stable friction on the rake face. Kelly and Cotterell (2002) investigated the effect of various types of lubrication (flood lubrication, MQL and compressed air) on the drilling of cast aluminium silicon alloys. MQL of vegetable oil gave lower feed forces and more superior finish at higher cutting speeds and feed rates. Costa et al. (2009) found that in drilling process, utilising vegetable oil as MQL lubricant produced smaller burr heights than that produced by mineral oil spayed using MQL technique.

Belluco and Chiffre (2004) reported that in drilling AISI 316L austenitic stainless steel, vegetable-based oils (rapeseed-ester-meadowfoam and rapeseed-ester blended oils) produced better results than the mineral reference oil, the best performance being 177 % increase in tool life and 7 % reduction in thrust force with respect to the commercial mineral oil. Castor oil-based cutting fluid formulated by Alves et al. (2013) was tested in grinding SAE 8640 steel with vitrified CBN wheel. Lower wheel wear and better surface finish was obtained in the tests using sulphonate castor oil with high water content as lubricant, as compared to semi-synthetic cutting fluid. The formulated fluid had no corrosion inhibiting characteristics when tested under corrosion test. Another workers (Ozcelik et al. 2011a) reported that vegetable-based cutting fluids (refined sunflower and canola oils) containing 8–12 % sulphur-based EP additive outperformed the semi-synthetic and mineral cutting fluids in turning AISI 304 austenite stainless steel. The vegetable-based cutting fluids resulted in better surface finish, lower flank and nose wear, and feed force. Ozcelik et al. (2011b) found that in drilling AISI 304 austenite steel, the refined sun-flower oil produced better surface finish than the semi-synthetic and mineral-based cutting fluids. Naves et al. (2013) utilised cutting fluid containing vegetable oil emulsifier with concentrations of 5 and 10 % at high pressure (10, 15 and 20 MPa) in machining ISI 316 austenitic stainless steel using coated cemented carbide tools. The flank wear, primarily subjected to adhesive wear, decreased with an increased in pressure and concentration of vegetable oil. Another researchers (Krishna et al. 2010) reported that among the two types of lubricants used (coconut oil and commercial SAE-40 oils, both containing nanoboric acid particles) the former lubricant produced lower flank wear, surface roughness and cutting temperature in machining AISI 1040 steel.

Liew and Ding (2008) and Liew (2010) found that in machining stavax® (modified AISI 420 stainless steel) with the hardness of 55 HRC using carbide end mills PVD coated with a single-layer TiAlN, several distinct stages of tool wear occurred; initial wear by delamination, attrition and abrasion, followed by cracking at the substrate and the formation of individual surface fracture at the cracks which would then enlarge and coalesce to form a large fracture surface. Compare to the flood lubrication (91 vol% water and 9 vol% SDBL oil), small quantity of mineral oil sprayed in mist form was more effective in reducing the coating delamination and delaying the occurrence of cracking and fracture. The effectiveness of mineral oil in suppressing coating delamination and delaying the occurrence of cracking and fracture could be enhanced by the presence of palm oil methyl ester (Liew et al. 2014). The mechanism by which the palm oil methyl ester suppressed these wear mechanisms could be explained by the results obtained in the four-ball tests which showed that the presence of palm oil methyl ester as additive in the mineral oil reduced the friction coefficient, severity of welding of the asperities and wear scar, and increased the critical load for welding to occur (refer to Sect. 3.2). This in turn reduced (i) the severity of the impact of the tool on the work material and (ii) the removal rate of the coating in the initial stage of machining, giving the tool substrate greater suppression of fatigue crack initiation.

References

Abou-El-Hossien KA, Yahya Z (2005) High-speed end-milling of AISI 304 stainless steels using new geometrically developed carbide inserts. J Mater Process Technol 162–163:596–602

Alves SM, Barros BS, Trajano MF, Ribeiro KSB, Moura E (2013) Tribological behaviour of vegetable oil-based lubricants with nanoparticles of oxides in boundary lubrication conditions. Tribol Int 65:28–36

Arndt M, Kacsich T (2003) Performance of new AlTiN coatings in dry and high speed cutting. Surf Coating Technol 163–164:674–680

Attanasio A, Gelfi M, Giardini C, Remino C (2006) Minimal quantity lubrication in turning: effect on tool wear. Wear 260:333–338

Belluco W, Chiffre LD (2004) Performance evaluation of vegetable-based oils in drilling austenitic stainless steel. J Mater Process Technol 148:171–176

Bhatia SM, Pandey PC, Shan HS (1978) Thermal cracking of carbide tools during intermittent cutting. Wear 51:201–211

Bhatia SM, Pandey PC, Shan HS (1980) The thermal condition of the tool cutting edge in intermittent cutting. Wear 61:21–30

Broniszewski K, Wozniak J, Czechowski K, Jaworska L, Olszyna A (2013) Al$_2$O$_3$-Mo cutting tools for machining hardened stainless steel. Wear 303:87–91

Chakraborty P, Asfour S, Cho A, Onar A, Lynn M (2008) Modelling tool wear progression by using mixed effects modelling technique when end milling AISI 4340 steel. J Mater Process Technol 205:190–202

Chou YK, Evans CJ (1997) Tool wear mechanism in continuous cutting of hardened tool steels. Wear 212:59–65

Costa ES, da Silva MB, Machado AR (2009) Burr produced on the drilling process as a function of tool wear and lubricant-coolant conditions. J Braz Soc Mech Sci Eng 31:57–63

Deng J, Song W, Zhang H, Yan P, Liu A (2011) Friction and wear behaviors of the carbide tools embedded with solid lubricants in sliding wear tests and in dry cutting processes. Wear 270:666–674

Deng J, Wu Z, Lian Y, Qi T, Cheng J (2012) Performance of carbide tools with textured rake-face filled with solid lubricants in dry cutting processes. Int J Refract Metal Hard Mater 30:164–172

Ding X, Liew WYH, Ngoi BKA, Gan JGK, Yeo SH (2002) Wear of CBN tools in ultra-precision machining of STAVAX. Tribol Lett 12:3–12

Dolinsek S, Sustarsic B, Kopac J (2001) Wear mechanism of cutting tool in high-speed cutting processes. Wear 250:349–356

Elkhabeery MM, Bailey JA (1984) Surface integrity in machining solution-treated and aged 2024-aluminium alloy, using natural and controlled contact length tools. part 1-unlubricated conditions. J Eng Mater Technol Trans ASME 106:152–160

Endrino JL, Fox-Rabinovich GS, Grey C (2006) Hard AlTiN, AlCrN PVD coatings for machining of austenite stainless steel. Surf Coat Technol 200:6840–6845

Erdemir A (2005) Review of engineered tribological interfaces for improved boundary lubrication. Tribol Int 38:249–256

Ezugwu EO, Bonney J (2005) Finish machining of nickel-base Inconel 718 alloy with coated carbide tool under conventional and high-pressure coolant supplies. Tribol Trans 48:76–81

Ezugwu EO, Olajire KA (2002) Evaluation of machining performance of martensitic stainless steel (JETHETE). Tribol Lett 12:183–187

Ghani JA, Choudhury IA, Masjuki HH (2004) Wear mechanism of TiN coated carbide and uncoated cermets tools at high cutting speed applications. J Mater Process Technol 153–154:1067–1073

Grill A (1993) Review of the tribology of diamond-like carbon. Wear 168:143–153

Gu J, Barber G, Tung S, Gu RJ (1999) Tool life and wear mechanism of uncoated and coated milling tool. Wear 225:273–284

Haque T, Morina A, Neville A (2010) Influence of friction modifier and antiwear additives on the tribological performance of a non-hydrogenated DLC coating. Surf Coat Technol 204:4001–4011

Itoigawa F, Childs THC, Nakamura T, Belluco W (2006) Effects and mechanisms in minimal quantity lubrication machining of an aluminum alloy. Wear 260:339–344

Kaminski J, Alvelid B (2000) Temperature reduction in the cutting zone in water-jet assisted turning. J Mater Process Technol 106:68–73

Kang MC, Kim KH, Shin SH, Jang SH, Park J, Kim C (2008) Effect of the minimum quantity lubrication in high-speed end-milling of AISI D2 cold-worked die Steel (62 HRC) by coated carbide tools. Surf Coat Technol 202:5621–5624

Kelly JF, Cotterell MG (2002) Minimal lubrication machining of aluminium alloys. J Mater Process Technol 120:327–334

Khan MMA, Mithu MAH, Dhar NR (2009) Effects of minimum quantity lubrication on turning ALSL 9310 alloy steel using vegetable oil-based cutting fluid. J Mater Process Technol 209:5573–5583

Kim SW, Lee DW, Kang MC, Kim JS (2001) Evaluation of machinability by cutting environments in high-speed milling of difficult-to-cut materials. J Mater Process Technol 111:256–260

Klocke F, Maßmann T, Gerschwiler K (2005) Combination of PVD tool coatings and biodegradable lubricants in metal forming and machining. Wear 259:1197–1206

Ko TJ, Kim HS, Chung BG (1999) Air-oil cooling method for turning of hardened material. Int J Adv Manuf Technol 15:470–477

Korkut I, Donertas MA (2007) The influence of feed rate and cutting speed on the cutting forces, surface roughness and tool-chip contact length during face milling. Mater Des 28:308–312

Krishna PV, Srikant RR, Rao DN (2010) Experimental investigation on the performance of nano-boric acid suspensions in SAE-40 and coconut oil during turning of AISI 1040 steel. Int J Mach Tools Manuf 50:911–916

Kubo T, Fujiwara S, Nanao H, Minami I, Mori S (2008) Boundary film formation from over-based calcium sulfonate additives during running in process of steel-DLC contact. Wear 265:461–467

Kumar AS, Durai AR, Sornakumar T (2006) The effect of tool wear on tool life of alumina-based ceramic cutting tools while machining hardened martensitic stainless steel. J Mater Process Technol 173:151–156

Liao YS, Lin HM (2007) Mechanism of minimum quantity lubrication in high-speed milling of hardened steel. Int J Mach Tools Manuf 47:1660–1666

Liao YS, Lin HM, Chen YC (2007) Feasibility study of the minimum quantity lubrication in high-speed end milling of NAK80 hardened steel by coated carbide tool. Int J Mach Tools Manuf 47:1667–1676

Liew WYH (2010) Low-speed milling of stainless steel with TiAlN single-layer and TiAlN/AlCrN nano-multilayer coated carbide tools under different lubrication conditions. Wear 269:617–631

Liew WYH, Ding X (2008) Wear progression of carbide tool in low-speed end milling of stainless steel. Wear 265:155–166

Liew WYH, Ngoi BKA, Lu YG (2003) Wear characteristics of PCBN tools in ultra-precision machining of stainless steel at low cutting speeds. Wear 254:265–277

Liew WYH, Lu YG, Ding X, Ngoi BKA, Yuan S (2004a) Performance of uncoated and coated carbide tools in the ultra-precision machining of stainless steel. Tribol Lett 17:851–857

Liew WYH, Yuan S, Ngoi BKA (2004b) Evaluation of machining performance of STAVAX with PCBN tools. Int J Adv Manuf Technol 23:11–19

Liew WYH, Dayou S, Dayou J, Siambun NJ, Ismail MAB (2014) The effectiveness of palm oil methyl ester as lubricant additive in milling and four-ball tests. Int J Surf Sci Eng 8:153–172

Lim CYH, Lim SC, Lee KS (1999) Wear of TiC-Carbide tools in dry turning. Wear 225–229:354–367

Liu N, Han C, Yang H, Xu Y, Shi M, Chao S, Xie F (2005) The milling performance of TiC-based cermet tools with TiN nanopowder addition against normalized medium carbon steel AISI 1045. Wear 258:1688–1695

Marinov V (1996) Experimental study on the abrasive wear in metal cutting. Wear 197:242–247

Minfray C, Fromentin G, Bierla A, Martin JM, Mogne TL (2014) The effect of an organic penta-sulfide EP additive in turning and milling operations. Wear 317:129–140

Mistry KK, Morina A, Neville A (2011) A tribochemical evaluation of a WC-DLC coating in EP lubrication conditions. Wear 271:1739–1744

Naves VTG, Da Silva MB, Da Silva FJ (2013) Evaluation of the effect of application of cutting fluid at high pressure on tool wear during turning operation of AISI 316 austenitic stainless steel. Wear 302:1201–1208

Nordin M, Sundström R, Selinder TI, Hogmark S (2000) Wear and failure mechanisms of multilayered PVD TiN/TaN Tools when milling austenite stainless steel. Surf Coat Technol 133–134:240–256

Nouari M, Ginting A (2006) Wear characteristics and performance of multi-layer CVD-coated alloyed carbide tool in dry end milling of titanium alloy. Surf Coat Technol 200:5663–5676

Nouari M, Molinari A (2005) Experimental verification of a diffusion tool wear model using a 42CrMo4 steel with an uncoated cemented tungsten carbide at various cutting speeds. Wear 259:1151–1159

Ozcelik B, Kuram E, Demirbas E, Sik E (2011a) Optimization of surface roughness in drilling using vegetable-based cutting oils developed from sunflower oil. Ind Lubr Tribol 63:271–276

Ozcelik B, Kuram E, Cetin MH, Demirbas E (2011b) Experimental investigations of vegetable based cutting fluids with extreme pressure during cutting of AISI 304L. Tribol Int 44:1864–1871

Rahim EA, Sasahara H (2011) A study of the effect of palm oil as MQL lubricant on high speed drilling of titanium alloys. Tribol Int 44:309–317

Rahman M, Kumar AS, Choudhury MR (2000) Identification of effective zones for high pressure coolant in milling. CIRP Annals-Manufact Technol 49:47–52

Rahman M, Kumar AS, Salam MU (2002) Experimental evaluation on the effect of minimal quantities of lubricant in milling. Int J Mach Tools Manuf 42:539–547

Reddy NSK, Rao PV (2006) Experimental investigation to study the effect of solid lubricants on cutting forces and surface quality in end milling. Int J Mach Tools Manuf 46:189–198

Rowe GW (1981) Lubrication in metal cutting and grinding. Phil Mag 43:567–585

Shaw MC (1984) Metal cutting Principles. Oxford University Press, Oxford

Sharif S, Yusof NM, Idris MH, Ahmad ZB, Sudin I, Ripin A and Mat Zin, MAH (2009) Feasibility study of using vegetable oil as a cutting lubricant through the use of minimum quantity lubrication during machining, Research VOT No. 78055, Department of Manufacturing and Industrial Engineering, Faculty of Mechanical Engineering, Universiti Teknologi Malaysia. http://eprints.utm.my/9729/1/78055.pdf

Shokrani A, Dhokia V, Newman ST (2012) Environmentally conscious machining of difficult-to-machine materials with regard to cutting fluids. Int J Mach Tools Manuf 57:83–101

Soderberg S, Hogmark S (1986) Wear mechanism and tool life of high speed steels related to microstructure. Wear 110:315–329

Sokovic M, Kopac J, Dobrzanki LA, Adamiak M (2004) Wear of PVD-coated solid carbide end mills in dry high-speed cutting. J Mater Process Technol 157–158:422–426

Stanford M, Lister PM, Kibble KA (2007) Investigation into the effect of cutting environment on tool life during the milling of a BS970-080A15 (En32b) low carbon steel. Wear 262:1496–1503

Su Y, He N, Li L, Li XL (2006) An experimental investigation of effects of cooling/lubrication conditions on tool wear in high-speed end milling of Ti–6Al–4V. Wear 261:760–766

Sun F, Li Z, Jiang D, Chen B (1998) Adhering wear mechanism of cemented carbide cutter in the intervallic cutting of stainless steel. Wear 214:79–82

Viera JM, Machado AR, Ezugwu EO (2001) Performance of cutting fluids during face milling of steels. J Mater Process Technol 116:244–251

Vivancos J, Luis CJ, Ortiz JA, Gonzalez HA (2005) Analysis of factors affecting the high-speed side milling of hardened die steels. J Mater Process Technol 162–163:696–701

Yallese MA, Chaoui K, Zeghib N, Boulanouar L, Rigal J (2009) Hard machining of hardened bearing steel using cubic boron nitride tool. J Mater Process Technol 209:1092–1104

Chapter 4
Utilisation of Environmental Friendly Gaseous and Vapour in Machining

Abstract Air, oxygen and nitrogen are examples of environmental friendly lubricants in the gaseous state. Even at low pressures, they have strong influences on the cutting performance. However, as the cutting speed increases, the effectiveness of the gas lubricants attenuated. This has conventionally been attributed to the reduction in interface penetration and thus of intimate contact and adhesion between chip and tool. This chapter also provides a review of the advanced techniques supplying gaseous and vapours capable of prolonging tool life at high cutting speeds.

Keywords Gaseous lubricant · Water vapour lubricant · Cryogenic machining · Chilled air machining · Environment friendly lubricants

4.1 Gaseous and Water Vapour as Lubricant in Machining

It has been recognised experimentally that two distinct mechanisms of friction coexist on the chip–tool interface. Doyle et al. (1979), Wallace and Boothroyd (1964) identified intimate contact between the chip and the tool over a region adjacent to the immediate vicinity of the cutting edge. This intimate contact is known as the sticking region. At a region some distance away from the cutting edge, sliding friction occurs. Frictional conditions between the chip and the rake face of the tool are very different from those in conventional sliding systems, in particular, the contact face of the chip is a freshly created clean metal surface and is subjected to very high levels of normal stress.

Over the part where sticking friction occurs, the normal stress σ rises steeply to a maximum value at the cutting edge (Fig. 4.1). Deformation occurs in the lower layers of the chip material and the real contact area approaches the apparent area. The shear stress is constant, equal to the shear stress of the chip k and independent

© The Author(s) 2015
W. Liew Yun Hsien, *Towards Green Lubrication in Machining*,
SpringerBriefs in Green Chemistry for Sustainability,
DOI 10.1007/978-981-287-266-1_4

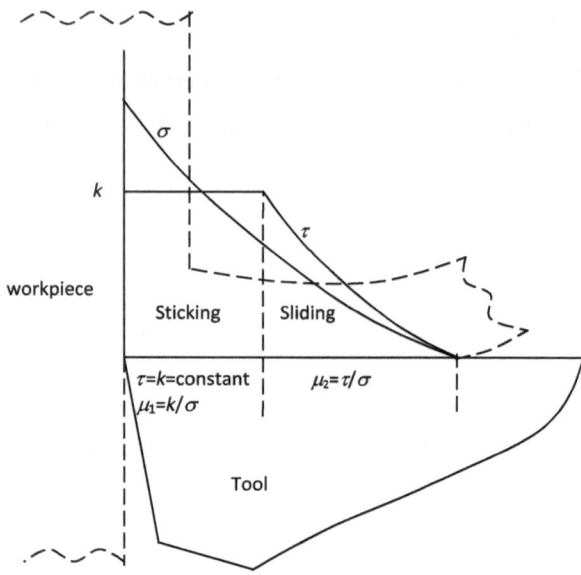

Fig. 4.1 Distribution of normal and shear stress on the tool rake face

of the normal stress σ. Over the part of this region, the coefficient of fiction μ is inversely proportional to the normal stress given by:

$$\mu_1 = \frac{k}{\sigma} \tag{4.1}$$

At some distance away from the cutting edge, sliding friction occurs in which the coefficient of friction is constant. In this region, relative motion between the chip and the tool occurs at the interface between the contacting asperities. Other evidence supporting the existence of this form of normal and shear stress distributions at the chip–tool interface has been presented by several researchers using photoelastic tools (Chandrasekaran and Kapoor 1965; Usui and Takeyama 1960). Over the part of this region, the shear stress τ decreases according to the law:

$$\mu_2 = \frac{\tau}{\sigma} \tag{4.2}$$

Wallace and Boothroyd (1964) concluded that in spite of the very low cutting speeds used in their investigation, penetration of lubricant vapour to the sticking region was improbable, since in this region no capillary network exists. Any lubricating activity was restricted to the sliding region. Recently, Wakabayashi et al. (1993) demonstrated that ethanol and carbon tetrachloride vapours were able to eliminate the sticking region and the transfer of chip material within the sliding region. This resulted in a marked reduction in the coefficient of friction and a superior surface finishes. Carbon tetrachloride is well known as an excellent lubricant in the cutting of many metals and this is attributed to the formation of weak chloride compounds at the interface. Furthermore, it is volatile and its molecules are small. These factors are expected to facilitate penetration to the zone of chip formation close to the cutting edge. Shaw (1951) observed that at low cutting speeds, carbon tetrachloride vapour was effective as the liquid lubricants. The results underline the potential importance of gaseous lubricants in metal cutting. However, the use of carbon tetrachloride had been discouraged due to the hazardous nature of the chemical. Many gaseous lubricants function as excellent lubricants only at low cutting speeds. At high cutting speeds, the condition of sliding contact between the tool and the chip become characteristics of unlubricated friction behaviour. The ineffectiveness of vapour lubricant at relatively higher speed has been attributed to the inaccessibility of the vapour to the contact zone and insufficient time for chemical reaction (Rowe and Smart 1964–1965).

Atmospheric oxygen provides a well-established example of an environmental friendly lubricant in the gaseous state; alone it can have a strong proven influence on cutting performance. Oxygen has a significant influence on cutting mild steel and this is attributed to the reduction of gross adhesion of the chip at the rake face (Rowe and Smart 1963, 1964–1965, 1966–1967; Rowe et al. 1976). However, apparently anomalous effects have been noted when machining non-ferrous materials such as aluminium and copper, where oxygen seems to increase the cutting forces (Williams 1975; Wright et al. 1979; Rowe and Smart 1966–1967; Doyle et al. 1979; Doyle and Horne 1980). This was explained by supposing that the tool would be subjected to the rubbing action of regions of oxide on the chip which might be sufficiently abrasive to remove the contaminant film on the tool resulting in high metallic contact (Williams and Stobbs 1979). An alternative interpretation could be that the relatively strong adsorbed oxide layer on the free or outer chip surface impedes dislocation movement at the shear plane, and counteracts the effect of reduced adhesion at the tool face (Rowe and Smart 1966–1967).

Liew (2004) carried out experiments to investigate the influence of air and pure oxygen on the machining behaviour of aluminium 2014A workpieces with different heat treatments at low speeds. High-speed-steel (HSS) tools were used in this study. Air and pure oxygen, which had been known to make the machining of non-ferrous material more difficult, were found to be effective in reducing the frictional force and the gross adhesion between the chip and the tool in the machining of precipitation hardened aluminium 2014A alloy. Figure 4.2 shows the variation of the coefficient of friction, μ, of Al T4 (aluminium 2014A-T4, i.e. aluminium 2014 which has been solution treated and naturally aged at room temperature) and Al T6 (aluminium 2014A-T6, i.e. aluminium 2014 which has been solution treated and artificially aged to the peak hardness) with cutting speeds in different atmospheres. Where, $\mu = F/W$. F and W are the frictional and normal forces acting parallel and normal to the tool rake, respectively (Fig. 4.3). The μ has been used to describe the state of rake face friction, and changes in its value is used to indicate the influence of lubricant on the cutting process.

In the machining of Al T4, the values of μ in air were consistently lower than those in vacuum at all cutting speeds. In the machining of Al T6, although the picture was clearly more complex, it was evident that in air, the values of μ were

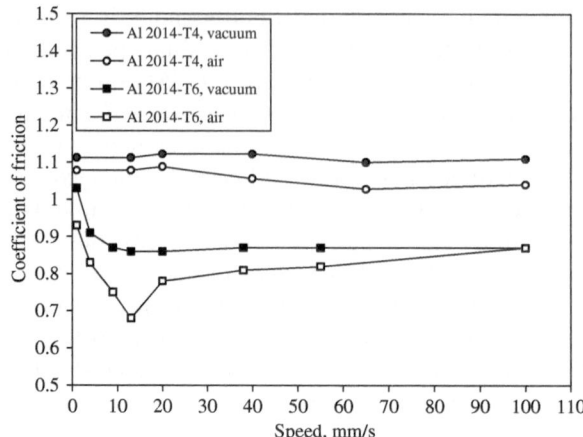

Fig. 4.2 Variation of rake face coefficient of friction with cutting speed for Al 2014-T4 and Al 2014-T6 cut in different atmospheres (Liew 2004)

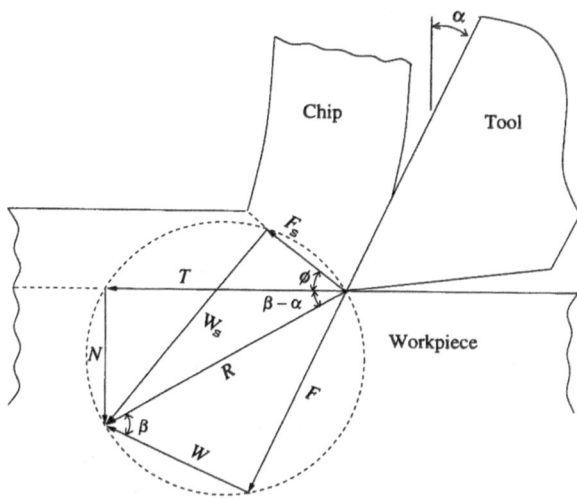

Fig. 4.3 Forces involved in orthogonal machining for a workpiece moving from left to right relative to the tool. The diagram shows the total force R exerted by the tool can be resolved into components, either F (frictional force) and W (normal force) or T (cutting force) and N (thrust force), respectively parallel and perpendicular to the rake face of the tool or the direction of motion. Thus, $F = T \sin \alpha + N \cos \alpha$ and $W = T \cos \alpha - N \sin \alpha$ where α is the rake angle. The mean angle of friction for the rake face contact β is related to the normal force W and frictional force F by the expression $\beta = \tan^{-1} \left(\frac{F}{W} \right)$ where $\beta = \tan^{-1} \mu$

lower than in vacuum. In machining Al T6 in vacuum, as the cutting speed was increased to 13 mm/s, μ decreased to a minimum after which it remained fairly steady even with further increase in the cutting speed. A similar, though more dramatic, trend was observed in air; at cutting speeds above 13 mm/s, μ increased

Table 4.1 Comparison of the coefficient of friction (COF) obtained in machining Al 2014-T6 in air, and 20 and 40 kPa pressure of pure oxygen at various speeds

Speed (mm/s)	COF in air	COF in 20 kPa oxygen	COF in 40 kPa oxygen
38	0.81	0.78	0.76
55	0.82	0.78	0.75
100	0.87	0.82	0.75

and gradually approached the value in vacuum. At 100 mm/s, the value of μ in air and in vacuum was essentially the same, suggesting that the effectiveness of air as lubricant ceased at this cutting speed. The influence of pure oxygen on the contact behaviour at the chip–tool interface at various cutting speeds was investigated. In machining Al T6 at 20 and 40 kPa pressure of pure oxygen, for the same cutting speed, μ is lower than that obtained in vacuum and air (Table 4.1).

Two distinct regions are evident on the tool contact surface produced at the low cutting speed of 1 mm/s (Fig. 4.4a, b); a dark region extending from the cutting edge some distance up the rake face, where there is little material transfer. Beyond this, a lighter region over which there is significant transfer of workpiece material. These regions are thought to correspond to zones 1 (sticking zone) and 2 (sliding zone) as defined by Doyle et al. (1979). Several researchers (Wallace and Boothroyd 1964; Chandrasekaran and Kapoor 1965; Usui and Takeyama 1960) have measured the normal and shear stress distributions at the chip–tool interface and found that the normal stress rises steeply to a maximum value at the cutting edge (Fig. 4.1). In zone 1 where the normal stress is high, deformation occurs in lower layers of the chip material and the real contact area approaches the apparent area. Over the remainder of the contact where zone 2 exists, relative motion between the chip and the tool occurs at the interface between the contacting asperities. In this zone, the real contact area is less than the apparent area. This has been explained in detail earlier in this chapter. SEM examination revealed that within zone 2, there were numerous transfer of chip material onto the cutting tool surface (Fig. 4.5). The severity of the transfer on the tool used to machine Al T6 in vacuum was much greater than that on the tool used to machine the same workpiece in air.

In machining Al T6, increasing the cutting speed from 1 to 13 mm/s caused an increase in the chip curl and thus a decrease in the overall contact area between the chip and the tool (Fig. 4.4). In vacuum, this coincided with a reduction in the friction force and μ. Since the normal force loading the chip onto the rake face was found to be invariant, the average normal stress acting on the chip surface and hence the density of true asperity contacts would increase as the overall contact area decreased. Therefore, as the contact area decreased, an increase in the extent of zone 1 and a reduction in the extent of zone 2 would take place. This effect was seen as the cutting speed was increased from 1 to 13 mm/s in vacuum (Fig. 4.4b, d, f). The rake faces of the tools used to cut Al T6 in vacuum at cutting speeds above 13 mm/s virtually showed no evidence of zone 2, i.e. only region corresponds to zone 1 was seen. In air, at a cutting speed of 100 mm/s, the extent of

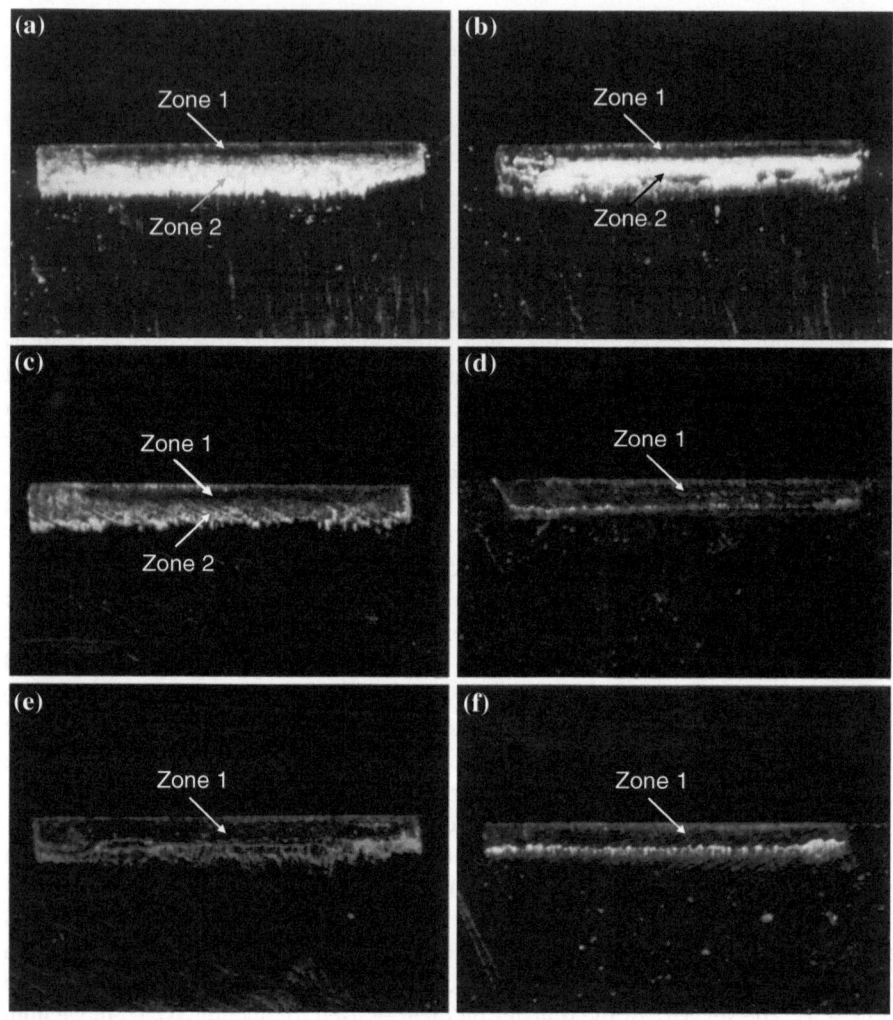

Fig. 4.4 Optical photographs of HSS tool tips used to machine Al 2014A-T6 for a distance of 0.5 m (Liew 2004). **a** in air at 1 mm/s ($\mu = 0.93$). **b** in vacuum at 1 mm/s ($\mu = 1.03$). **c** in air at 13 mm/s ($\mu = 0.69$). **d** in vacuum at 13 mm/s ($\mu = 0.86$). **e** in air at 100 mm/s ($\mu = 0.87$). **f** in vacuum at 100 mm/s ($\mu = 0.87$)

zone 1 was the same as that at the cutting speeds of 13 and 100 mm/s in vacuum (Fig. 4.4e, f). These observations were consistent with the measured values of μ which were found to be essentially the same. The light region appeared on the rake face of the tool used to machine Al T6 at 13 mm/s in air indicated that air had successfully penetrated into the intimate contact at the chip–tool interface (Fig. 4.4c). The frictional behaviour over the lighter region was similar to that of zone 2 where relative motion between the contacting asperities of the chip and the tool (sliding friction) occurs. Therefore, the overall frictional behaviour in air was such as to

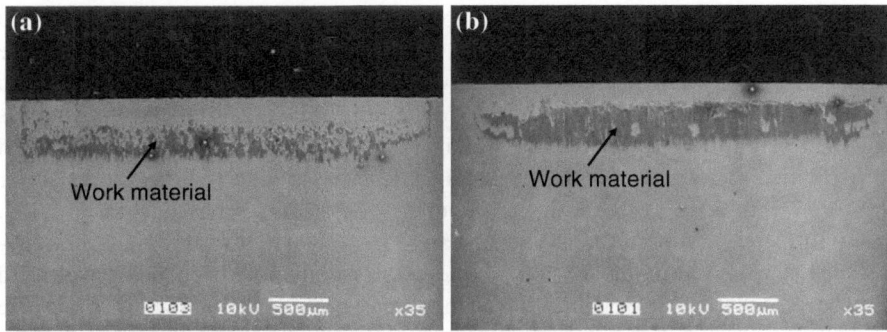

Fig. 4.5 SEM of HSS tool tips used to machine Al 2014A-T6 at 1 mm/s for a distance of 0.5 m. Lesser amount of work material adhering onto the tool used to machine the workpiece in air (Liew 2004). **a** In air. **b** In vacuum

give a lower frictional force than that in vacuum where deformation at the chip–tool interface occurred mostly in the lower layers of the chip material.

The degree of transfer onto the rake face of the tool used to machine Al T6 at an oxygen pressure of 20 kPa was significantly less than that observed on the rake face of the tool used to machine the alloy in vacuum. This result indicates that oxygen in atmospheric air plays an important role in reducing the gross transfer of the chip material onto the tool. At 20 kPa of pure oxygen, the values of μ were lower than those obtained at the same partial pressure of oxygen in atmospheric air. This could be due to the presence of nitrogen in air reacting with aluminium to form nitride. Blouet and Courtel (1975) investigated the influence of oxygen on the frictional behaviour of aluminium alloys sliding on steel. They found that oxygen could reduce, or prevent, the transfer of aluminium onto the steel surface by virtue of its ability to form a protective oxide film.

In the past, most investigations on the influence of gaseous had been carried out at or below atmospheric pressures. As the cutting speed was increased, the effectiveness of the gas lubricants decreased significantly and eventually ceased. Gaseous are poor coolants and therefore not able to dissipate heat generated at high cutting speeds. Recent research works focused on improving the effectiveness of gaseous lubricants at high cutting speeds. Liu et al. (2007) demonstrated the effectiveness of water vapour, gas (carbon dioxide and oxygen), and mixture of vapour and gas as lubricants in machining steel up to the speed of 117.6 m/min. The best lubricant (measured in terms of reduction in cutting force and temperature, and increase in tool life) was found to be governed by the cutting speed. At the lower cutting speeds of 60 and 67.2 m/min, mixture of gas and vapour was more effective in suppressing in flank wear compared to carbon and oxygen gas. The vapour acted as both lubricant and coolant while gas primarily acted as lubricant. However at the cutting speeds of 84 and 117.6 m/min, oxygen and carbon dioxide gas produced lower flank wear.

More recently, Wu and Han (2013) used high-speed-steel and cemented carbide bits in drilling titanium alloy Ti6Al4V under dry and lubrication conditions up to

12.09 m/min utilising water vapour (supplied at 0.18–0.26 MPa) as coolant and lubricant. When water vapour was used, the torque, thrust and temperature were reduced by 5–25 %, and the tool wear was reduced by 10–80 % in comparison to drilling under dry condition and wet condition utilising oil water emulsion as lubricant. The effectiveness of the water vapour as lubricant could be attributed to its ability to penetrate to the tool–chip interface through the capillaries (refer to Sect. 1.3) at a faster rate to form a boundary lubrication layer (Liu et al. 2005, 2009). It has been reported that vapour can react with steel to form interfacial layers such as iron hydroxide and ferri-oxide-hydrates effective in reducing wear and friction (Liew 2006). Moisture can also adsorb physically on the worn surface and act in a protective manner so as to prevent direct contact between the surfaces (Goto and Buckley 1985). In turning AISI 316 stainless steel, in comparison to flood lubrication, CO_2 sprayed at 3 g/s was capable of reducing the cutting temperature up to 35 %, and improving the surface finish by 4–52 % with reduced tool wear and better chip breakability over a wide range of speeds of between 41–145 m/min (Jerold and Kumar 2012).

4.2 Cryogenic Machining

In cryogenic machining, refrigerated compressed gas is jetted into the cutting zone to allow machining to be carried out at very low temperatures, typically lower than 120 K. It is another alternative method to increase tool life. The cooling effect is particularly useful in machining materials with low thermal conductivity such as nickel based alloys that could result in excessive tool wear due to high cutting temperature. In order to compensate the lack of lubrication, a very small amount of cutting oil may be added in the cold gas stream. Liquefied nitrogen, carbon dioxide and helium are the typical cryogenic coolants used in this lubrication technique. Cryogenic coolant can significantly reduce the temperature chip–tool interface and thus the chemical reaction between the tool and the chip (Hong and Ding 2001; Bermingham et al. 2011). This allows machining to be carried out at higher speeds with little concern on thermal induced wear.

The cryogenic cooling can bring about beneficial effects on tool life and surface finish up to 550 m/min (Stanford et al. 2007). Stanford et al. (2007) reported that nitrogen gas resulted in the lowest flank wear and thermal crack compared to compressed air in high-speed milling of low carbon steel. Applying liquid nitrogen jet during machining steel was found to improve the dimensional accuracy of the machined parts, and reduce surface roughness and tool wear through the reduction in the cutting temperature and favourable change in the chip–tool and work–tool interactions at the cutting speeds of 60–150 m/min (Dhar et al. 2002). Dhananchezian and Kumar (2011) reported that in machining Ti–6Al–4V at 27–97 m/min with cryogenic cooling, the cutting temperature, cutting force and flank wear were 61–66, 35–42 and 27–39 %, respectively, lower than those produced under flood lubrication. In milling hardened AISI D3 tool steel at 125 m/min, cryogenic cooling was capable of further reducing the cutting temperature and mean cutting forces by 26–35 and 22–39 %, respectively, over those produced by flood lubrication (Ravi and Kumar 2012).

In machining Inconel 718 where intense heat was generated, this lubrication technique produced less tool wear and better surface finish compared to the results obtained in dry and MQL machining. However, at low speed, MQL produced better results (Kaynak 2014). In machining stainless steel where tool wear was governed by attrition, abrasion and micro-cracking, tool life produced by cryogenic coolant was more than four times higher than that produced by conventional coolant (Khan and Ahmed 2008). The cooling effect can result in an increase in the hardness of the work material but a reduction in the coefficient of friction resulting in improved machining condition (Hong 2006; Hong et al. 2001).

4.3 Chilled Air in Machining

This technique, one of the cleanest and most environmentally friendly method of cooling, used compressed and chilled air to cool the chip–tool interface. Rahman et al. (2003) investigated the effect of chilled air and conventional flood coolant in end milling of steel. They found that chilled air supplied at 0.5 MPa resulted in lower wear only at low cutting speed. However, Kim et al. (2001) had reported beneficial effects of chilled air in high-speed milling of steel (up to 210 m/min). They found that compressed and chilled air resulted in better tool life than flood coolant and dry machining. The tool life was 3.5 times higher than under flood coolant and 2 times higher than under dry machining. Refrigerated compressed air also generated lower thermal fatigue wear compared to flood coolant in high-speed milling of steel. Since the cutting heat was not removed as much as under flood coolant, the thermal cracking was less likely to form, resulting in longer tool life. Su et al. (2007) investigated the effect of chilled air (supplied at 0.6 MPa) on the tool wear, surface finish and chip shape in high-speed milling of AISI D2 cold work tool steel at 76 m/min. They found that chilled air resulted longer tool life but slightly higher surface roughness than dry cutting. Liu and Chou (2007) had studied the effects of chilled air in turning silicon–aluminium at 186 and 300 m/min, and found that chilled air reduced the flank wear and temperature at the chip–tool interface up to 20 and 7 %, respectively. It appeared that the effect of the chilled air on the tool wear was dependent on the machining parameter used with the combination of high speed and low feed being the most effective. The lubrication effect of the chilled air can be improved by adding a small amount of oil into the chilled air (Nyugan and Zhang 2003).

Kim et al. (2001) found that compressed chilled air provided longer tool life than flood coolant. While chilled air increased the tool life in machining hardened steel, this beneficial effect was not seen in high-speed machining of Inconel. This could be attributed to the poor thermal conductivity of the air where the heat generated greatly surpassed the cooling effect brought about by the chilled air. Several researchers integrated chilled air and MQL to produce the best machining results measured in terms of tool life and surface finish (Su et al. 2007; Yuan et al. 2011; Su et al. 2006). In particular, Su et al. (2007) showed that chilled MQL produced longer tool life and better surface finish than chilled air, and these lubrication techniques produced better results than dry machining.

References

Bermingham M, Kirsch J, Sun S, Palanisamy S, Dargusch MS (2011) New observations on tool life, cutting forces and chip morphology in cryogenic machining Ti-6Al-4V. Int J Mach Tools Manuf 51:500–511

Blouet J, Courtel R (1975) Phases of wear of the aluminum/steel couple in lubricated friction. Wear 34:109–125

Chandrasekaran H, Kapoor DV (1965) Photoelastic analysis of tool-chip interface stresses. J Eng Ind Trans ASME 87:495–502

Dhananchezian M, Kumar MP (2011) Cryogenic turning of the Ti–6Al–4V alloy with modified cutting tool inserts. Cryogenics 51:34–40

Dhar NR, Paul S, Chattopadhyay AB (2002) The influence of cryogenic cooling on tool wear, dimensional accuracy and surface finish in turning AISI 1040 and E4340C steels. Wear 249:932–942

Doyle ED, Horne JG (1980) Adhesion in metal cutting: anomalies associated with oxygen. Wear 60:383–391

Doyle ED, Horne JG, Tabor D (1979) Frictional interactions between chip and rake face in continuous chip formation. Proc Roy Soc Lond 366(173):183

Goto H, Buckley DH (1985) The influence of water vapour in air on the friction behaviour of pure metals during fretting. Tribol Int 18:237–245

Hong S (2006) Lubrication mechanisms of LN2 in ecological cryogenic machining. Mach Sci Technol 209:133–155

Hong SY, Ding Y (2001) Cooling approaches and cutting temperatures in cryogenic machining of Ti-6Al-4V. Int J Mach Tools Manuf 41:1417–1437

Hong SY, Ding Y, Jeong W (2001) Friction and cutting forces in cryogenic machining of Ti-6Al-4V. Int J Mach Tools Manuf 41:2271–2285

Jerold BD, Kumar MP (2012) Machining of AISI 316 stainless steel under carbon-di-oxide cooling. Mater Manuf Processes 27:1059–1065

Kaynak Y (2014) Evaluation of machining performance in cryogenic machining of inconel 718 and comparison with dry and MQL machining. Tribol Lett 72:919–933

Khan AA, Ahmed MI (2008) Improving tool life using cryogenic cooling. J Mater Process Technol 196:149–154

Kim SW, Lee DW, Kang MC, Kim JS (2001) Evaluation of machinability by cutting environments in high-speed milling of difficult-to-cut materials. J Mater Process Technol 111:256–260

Liew WYH (2004) The effect of air in the machining of aluminium alloy. Tribol Lett 17:41–49

Liew WYH (2006) The effect of relative humidity on the unlubricated wear of metals. Wear 260:720–727

Liu J, Chou KY (2007) On temperature and tool, wear in machining hypereutectic Al-Si alloys with vortex-tube cooling. Int J Mach Tools Manuf 47:635–645

Liu J, Han R, Sun Y (2005) Research on experiments and action mechanism with water vapour as coolant and lubricant in green cutting. Int J Mach Tools Manuf 45:687–694

Liu J, Han R, Zhang L, Guo H (2007) Study on lubricating characteristic and tool wear with water vapor as coolant and lubricant in green cutting. Wear 262:442–452

Liu JY, Liu HP, Han RD, Wang Y (2009) The study on lubrication action with water vapor as coolant and lubricant in cutting ANSI 304 stainless steel. J Mach Tools Manuf 49:260–269

Nyugan T, Zhang LC (2003) An assessment of the applicability of cold air and oil mist in surface grinding. J Mater Process Technol 140:224–230

Rahman M, Kumar AS, Salam MU, Ling MS (2003) Effect of chilled air on machining performance in end milling. Int J Adv Manuf Technol 21:787–795

Ravi S, Kumar MP (2012) Experimental investigation of cryogenic cooling in milling of AISI D3 tool steel. Mater Manuf Processes 27:1017–1021

Rowe GW, Smart EF (1963) The importance of oxygen in dry machining of metal on a lathe. J Appl Phys 14:924–926

Rowe GW, Smart EF (1964–1965) Experiments on lubrication breakdown in friction tests in cutting of metal on a lathe. Proc Inst Mech Eng (Part 3J) 179:229–241

Rowe GW, Smart EF (1966–1967) Vapour lubrication in friction and low speed metal cutting. Proc Inst Mech Eng 181:248–257

Rowe GW, Smart EF, Tripathi KC (1976) Surface adsorption effects in metal cutting and grinding. ASLE Trans 20:347–353

Shaw MC (1951) The metal cutting process as a means of studying the properties of extreme pressure lubricants. Ann NY Acad Sci 53:962–978

Stanford M, Lister PM, Kibble KA (2007) Investigation into the effect of cutting environment on tool life during the milling of a BS970-080A15 (En32b) low carbon steel. Wear 262:1496–1503

Su Y, He N, Li L, Li X (2006) An experimental investigation of effects of cooling/lubrication conditions on tool wear in high-speed end milling of Ti-6Al-4V. Wear 261:760–766

Su Y, He N, Li L, Iqbal A, Xiao MH, Xu S, Qiu BG (2007) Refrigerated cooling air cutting of difficult-to-cut materials. Int J Mach Tools Manuf 47:927–933

Usui E, Takeyama H (1960) A photoelastic analysis of machining stresses. J Eng Ind 82:303–308

Wakabayashi T, Williams JA, Hutchings IM (1993) The action of gaseous lubricants in the orthogonal machining of an aluminium alloy by titanium nitride coated tools. Surf Coat Technol 57:183–189

Wallace PW, Boothroyd G (1964) Tool forces and tool-chip friction in orthogonal machining. J Mech Eng Sci 6:74–87

Williams JA (1975) The role of lubricants in metal cutting, PhD thesis, University of Cambridge

Williams JA, Stobbs WM (1979) Changes in mode of chip formation as function of presence of oxygen. Metals Technol 6:424–432

Wright PK, Horne JG, Tabor D (1979) Boundary conditions at the chip-tool interface in machining: comparisons between seizure and sliding friction. Wear 54:317–390

Wu J, Han RD (2013) Research on experiments with water vapour as coolant and lubricant in drilling TiAl4V. Ind Lubr Tribol 65:50–60

Yuan SM, Yan LT, Liu WD, Liu Q (2011) Effects of cooling air temperature on cryogenic machining of Ti-6Al-4V alloy. J Mater Process Technol 211:356–362

Chapter 5
Conclusions

The environmental and toxicity issues of commercial additives and lubricants as well as their rising cost due to the depletion of world fuel reserves led to renewed interest in the development of alternative resources as lubricants and additives. Vegetable oils are potential alternative resources because of their environment-friendly, non-toxic and readily biodegradable nature. Despite their poor oxidation stability and corrosion protection, researches have shown that vegetable oils and their derivatives can act as effective lubricants and additives. In some cases, vegetable oils produced lower coefficient of friction and wear than the commercial lubricants. Additives derived from vegetable oils can be used to enhance the performance of commercial lubricants. On the other hand, the effectiveness of vegetable oils as lubricants can be enhanced by the presence of commercial additives. Various types of vegetable oils were found to produce better results (measured in terms of reduction in the cutting forces, temperature and tool wear) than the conventional cutting fluids in machining. More recently, some researchers have demonstrated that vegetable oils can be utilised as emulsifiers and additives in commercial cutting fluids. One of the most significant attribute of successful cutting lubricants is their ability to penetrate into the chip–tool interface to form an effective boundary lubricant. It is postulated that petroleum-based oils are, for the most part, non-polar, whereas triglycerides of vegetable oils have more polar groups, thus possess more sites for additive elements to react and adsorb with metal surfaces to form an effective boundary lubricant. Many vegetable oils, proven to be effective lubricants and additives in the fundamental research works carried out on tribotesters, are yet to be tested in machining. While it has been reported that vegetable oils can act as effective lubricants in machining materials like stainless steel and titanium, the limitations and effectiveness of these lubricants are yet to be tested on difficult-to-machine materials such as Inconel. Not many research works have studied the nature of the lubricating films formed by vegetable oils used as lubricants and additives on the tool and the chip surfaces. Currently, there are no clear guidelines for selection of vegetable oils as lubricants in machining. A more rigorous tests are required to produce more data for construction of guidelines.

© The Author(s) 2015
W. Liew Yun Hsien, *Towards Green Lubrication in Machining*,
SpringerBriefs in Green Chemistry for Sustainability,
DOI 10.1007/978-981-287-266-1_5

Minimum quantity lubrication (MQL) and mist lubrication are the alternative lubrication methods designed to reduce the dependency on commercial lubricants. Spraying small quantity of lubricants using MQL technique and in mist form not only reduce the quantity of lubricant used in machining but also can result in greater reduction in the cutting forces, longer tool life and better surface finish, as compared to flood lubrication. MQL does generate some amount of mist, whereas mist lubrication generates large amount of mist. Filtering systems are required to manage the resulting fine mist which posed serious health hazards such as irritation and respiratory problems. Many researchers have reported promising results in machining tests utilizing vegetable oils as MQL lubricant.

One way to eliminate totally the usage of lubricant is by carrying out the machining under dry condition. While some alloys can be machined to superior finish and form under dry condition, many other alloys showed poor machinability in the absence of lubricant. There is a limited range of speed and materials that dry machining results in acceptable surface finish and tool life. In particular, the intense heat generated in high-speed machining under dry condition can cause severe tool wear and surface deterioration. Therefore, coolant is needed to dissipate the heat from the cutting zone. Machining soft materials in the absence of lubricant is likely to produce poor surface finish due to the formation of BUE. Deposition of solid lubricants such as graphite and molybdenum disulphide on the cutting tools can result in the formation of self-lubricating film at the tool-chip interface and a marked improvement in tool life and surface finish in dry machining.

Gaseous such as air, oxygen and nitrogen can be used as lubricants in machining. However, the effectiveness of the gas lubricants reduced significantly as the cutting speed is increased. This has conventionally been attributed to the reduction in interface penetration and thus growth of intimate contact and adhesion between chip and tool. Recent works have shown that gaseous and vapours supplied at high pressure are capable of prolonging the tool life at high cutting speeds. Gaseous have high ability to penetrate into the chip–tool interface. However, they are poor coolants and therefore not able to dissipate heat generated at high cutting speeds. This deficiency can be overcome by adding vapour into the gaseous. Chilled air and refrigerated compressed gas jetted into the cutting zone are capable of significantly reducing the cutting temperature, leading to reduced tool wear. In general, these lubrication techniques are more effective than other lubrication methods in machining at high speeds where excessive heat is generated. The lubrication effect of the chilled air can be improved by adding a small amount of oil. Chilled air and MQL can be integrated to produce better machining results.